Grade One Science

Aligned to Alberta Curriculum

Written by Tracy Bellaire and Andrew Gilchrist

This teacher resource has been designed to give students an understanding and appreciation of the 5 topics in Grade 1 Science Alberta Curriculum. Those 5 topics are Creating Colour, Seasonal Changes, Building Things, Senses and the Needs of Animals and Plants. The lessons are designed to involve tactile participation and knowledge application while providing opportunities to connect ideas between topics and school subjects.

TRACY BELLAIRE is an experienced teacher who continues to be involved in various levels of education in her role as Differentiated Learning Resource Teacher in an elementary school in Ontario. She enjoys creating educational materials for all types of learners, and providing tools for teachers to further develop their skill set in the classroom. She hopes that these lessons help all to discover their love of science!

ANDREW GILCHRIST taught primary and junior classes for the Hastings Prince Edward District School Board in Ontario. He earned his teaching degree with a Teaching Subject qualification in History. He recently co-authored a series of mathematics programs that focus on fluency skills with basic math operations and fractions. The series, used for children with special education needs, was used for a study with University of Ottawa. Students using the program demonstrated achievements in fluency at twice the rate of regular in-class instruction, allowing them to catch up with peers.

Copyright © On The Mark Press 2016

Some material appearing in this book has been used in other published works, such as Physical Science Grade 1 (OTM2144), Earth and Space Science Grade 1 (OTM2152) and Life Sciences Grade 1 (OTM2160).

This publication may be reproduced under licence from Access Copyright, or with the express written permission of On The Mark Press, or as permitted by law. All rights are otherwise reserved, and no part of this publication may be reproduced, stored in a retrieval system, or transmitted in any form or by any means, electronic, mechanical, photocopying, scanning, recording or otherwise, except as specifically authorized.

All Rights Reserved.
Printed in Canada.

Published in Canada by:
On The Mark Press
15 Dairy Avenue, Napanee, Ontario, K7R 1M4
www.onthemarkpress.com

OTM-2170   ISBN: 978-1-4877-0198-7 © On The Mark Press

AT A GLANCE

General Outcomes

1.1 Bring focus to investigative activities, based on their own questions and those of others. Students will ask questions that lead to exploration and predict what they think will happen or what they might find.

1.2 Describe materials and objects that have been observed and manipulated, and identify what was done and found out. Students will identify materials used, describe steps followed and observe the results of their activities and the activities of others.

1.3 Construct, with guidance, an object that achieves a given purpose, using materials that are provided. Students will identify a problem or task, attempt to complete the task and engage in all parts of the task, identify processes by which an object was made and identify or show how a product might be used.

Attitudes

1.4 Demonstrate positive attitudes to the study of science and to the application of science. Students will show growth in acquiring and applying curiosity, inventiveness, perseverance, appreciation of the value of experience and observation, a willingness to work with others, a sense of responsibility for actions taken and respect for living things and environments with a commitment for their care.

Topic A: Creating Colour

1.5 Identify and evaluate methods for creating colour and for applying colours to different materials. Students will identify colours in natural and manufactured objects, compare and contrast colours using terms such as lighter, darker, more blue, brighter than, arrange and order a group of coloured objects based on a given colour criterion, predict and describe colour changes when mixing a primary colour with white or black, distinguish between colours that are transparent and opaque, compare the effects of thickness on paint, recognize how a thick layer of paint can be partly transparent, compare how paint adheres to different surfaces, demonstrate how a colour can be extracted from certain materials and applied to others and demonstrate at least one way to separate sunlight into component colours.

Topic B: Seasonal Changes

1.6 Describe seasonal changes and interpret the effects of season changes on living things. Students will describe changes in the seasonal cycle using changes in sunlight and weather, identify examples of plant and animal changes that occur on a seasonal basis involving form, appearance, location or movement of habitats and changes in activity, identify human preparations for seasonal change, identify activities done on a seasonal basis and record observable seasonal changes over a period of time.

Topic C: Building Things

1.7 Construct objects and models of objects, using a variety of different materials. Students will select appropriate materials based on what they will construct – model buildings, objects, toys, wind or water related artifacts and identify the overall purpose for each model and artifact constructed.

1.8 Identify the purpose of different components in a personally constructed object or model, and identify corresponding components in a related object or model. Students will identify component parts of constructed objects and describe the purpose of the parts, compare objects that have been constructed for the same purpose, identify similarities and differences between parts and recognize that products are often developed for specific purposes.

Topic D: Senses

1.9 Use the senses to make general and specific observations, and communicate observations orally and by producing captioned pictures. Students will identify the human senses and choose what senses to use to describe objects, materials, living things and environments.

1.10 Describe the role of the human senses and the senses of others living things, in enabling perception. Students will identify how we use the senses to interpret the world for our safety and quality of life, recognize the limitations of our senses, recognize that other living things have senses, identify how animals use their senses, describe how people adapt to limited or loss of sensory abilities and describe ways to take care of our sensory organs.

Topic E: Needs of Animal and Plants

1.11 Describe some of the common living things, and identify needs of those living things. Students will contrast living and non-living things, observe, describe, compare and classify living things based on visible characteristics, identify ways in which living things are valued, identify examples of plants and animals normally considered under human care (domesticated) and independent of human care (wild), identify requirements to main life for animals and for plants, identify ways plants depend on soil, recognize how some plants and animals adapted to extreme conditions to meet their basic needs and give examples of ways in which animals depend on plants and ways in which plants depend on animals.

Taken from the Alberta Education Grade 1 Science Curriculum.

TABLE OF CONTENTS

AT A GLANCE 2
TEACHER ASSESSMENT RUBRIC 5
STUDENT SELF-ASSESSMENT RUBRIC 6
INTRODUCTION 7

TOPIC A: CREATING COLOUR
Identifying Colours 8
Primary and Secondary Colours 10
Applying Colours 15

TOPIC B: SEASONAL CHANGES
Seasonal Changes 20
Aboriginal Activities 33
The Heat of the Sun 39
Animal Adaptations 52
Plant Adaptations 61

TOPIC C: BUILDING THINGS
Constructing a Birdfeeder 68
Building a Pop-Up Creature Card and Bulding a Toy Animal 71
Building a Water Wheel and a Boat 78

TOPIC D: SENSES
Explore Your Senses 84
Applying Your Senses 93
Animal Senses 103
Protecting the Senses 109

TOPIC E: NEEDS OF ANIMALS AND PLANTS
The Animal World 121
The Plant World 128
Needs Intertwined 148

Photo credits: Page 89 Guitar: Chromakey / Shutterstock.com

Teacher Notes

Teacher Assessment Rubric

Student's Name: _____ Date: _____

Success Criteria	Level 1	Level 2	Level 3	Level 4
Knowledge and Understanding Content				
Demonstrate an understanding of the concepts, ideas, terminology definitions, procedures and the safe use of equipment and materials	Demonstrates limited knowledge and understanding of the content	Demonstrates some knowledge and understanding of the content	Demonstrates considerable knowledge and understanding of the content	Demonstrates thorough knowledge and understanding of the content
Thinking Skills and Investigation Process				
Develop hypothesis, formulate questions, select strategies, plan an investigation	Uses planning and critical thinking skills with limited effectiveness	Uses planning and critical thinking skills with some effectiveness	Uses planning and critical thinking skills with considerable effectiveness	Uses planning and critical thinking skills with a high degree of effectiveness
Gather and record data, and make observations, using safety equipment	Uses investigative processing skills with limited effectiveness	Uses investigative processing skills with some effectiveness	Uses investigative processing skills with considerable effectiveness	Uses investigative processing skills with a high degree of effectiveness
Communication				
Organize and communicate ideas and information in oral, visual, and/or written forms	Organizes and communicates ideas and information with limited effectiveness	Organizes and communicates ideas and information with some effectiveness	Organizes and communicates ideas and information with considerable effectiveness	Organizes and communicates ideas and information with a high degree of effectiveness
Uses science and technology vocabulary in the communication of ideas and information	Uses vocabulary and terminology with limited effectiveness	Uses vocabulary and terminology with some effectiveness	Uses vocabulary and terminology with considerable effectiveness	Uses vocabulary and terminology with a high degree of effectiveness
Application of Knowledge and Skills to Society and Environment				
Apply knowledge and skills to make connections between science and technology to society and the environment	Makes connections with limited effectiveness	Makes connections with some effectiveness	Makes connections with considerable effectiveness	Makes connections with a high degree of effectiveness
Propose action plans to address problems relating to science and technology, society, and environment	Proposes action plans with limited effectiveness	Proposes action plans with some effectiveness	Proposes action plans with considerable effectiveness	Proposes action plans with a high degree of effectiveness

OTM-2170 ISBN: 978-1-4877-0198-7 © On The Mark Press

STUDENT SELF-ASSESSMENT RUBRIC

Name: _____ Date: _____

Put a checkmark in the box that best **described you.**

Expectations	Need to do Better	Sometimes	Almost Always	Always
I am a good listener.				
I followed the directions.				
I stayed on task and finished on time.				
I remembered safety.				
My writing is neat.				
My pictures are neat and coloured.				
I reported the results of my experiment.				
I discussed the results of my experiment.				
I know what I am good at.				
I know what I need to work on.				

I liked _____

_____

I learned _____

_____

I want to learn more about _____

_____

.... Teacher Notes

INTRODUCTION

The lessons and experiments in this book fall under 5 main topics that relate to the Alberta curriculum for Grade 1 Science – Topic A: Creating Colours, Topic B: Seasonal Changes, Topic C: Building Things, Topic D: Senses and Topic E: Needs of Animals and Plants. In each lesson you will find teacher notes designed to provide you guidance with the learning expectations, the success criteria, materials needed, a lesson outline, as well as provide some insight on what results to expect when the experiments are conducted. Suggestions for differentiation or accommodation are also included so that all students can be successful in the learning environment.

Throughout the experiments, the scientific method is used. The scientific method is an investigative process which follows five steps to guide students to discover if evidence supports a hypothesis.

1. **Consider a question to investigate.** For each experiment, a question is provided for students to consider. For example, "Does direct sunlight affect water temperature?"

2. **Predict what you think will happen.** A hypothesis is an educated guess about the answer to the question being investigated. For example, "I believe that direct sunlight warms up the temperature of water." A group discussion is ideal at this point.

3. **Create a plan or procedure to investigate the hypothesis.** The plan will include a list of materials and a list of steps to follow. It forms the "experiment."

4. **Record all the observations of the investigation.** Results may be recorded in written, table, or picture form.

5. **Draw a conclusion.** Do the results support the hypothesis? Encourage students to share their conclusions with their classmates, or in a large group discussion format.

The experiments in this book fall under seventeen topics that relate to four aspects of physical science: Materials, Objects, and Building Things; Energy in Our Lives; Force and Motion; and Creating Colour. In each section you will find teacher notes designed to provide you guidance with the learning intention, the success criteria, materials needed, a lesson outline, as well as provide some insight on what results to expect when the experiments are conducted. Suggestions for differentiation are also included so that all students can be successful in the learning environment.

ASSESSMENT & EVALUATION:

Students can complete the Student Self-Assessment Rubric in order to determine their own strengths and areas for improvement. Assessment can be determined by observation of student participation in the investigation process. The classroom teacher can refer to the Teacher Assessment Rubric and complete it for each student to determine if the success criteria outlined in the lesson plan has been achieved. Determining an overall level of success for evaluation purposes can be done by viewing each student's rubric to see what level of achievement predominantly appears throughout the rubric.

ASSESSMENT & DIFFERENTIATION

Teachers may choose to start the unit with a student duotang which the students can keep all of their work in. The lessons have been designed to work independently or as a unit of study, with high interest and low vocabulary as leading goals. The Glossary of Terms after the Table of Contents can be copied for students and used as a reference, a study game or a "living document" where students add new terms for their own notes.

MEETING YOUR STUDENTS' NEEDS:

Depending on the needs of the students in your class, the teacher may want to scan any Teacher Notes into a digital format. By doing this, no matter the reading abilities of your students, they will be able to access the information of the text. When appropriate, teachers are also encouraged to allow students to collaborate on as many activities possible. This allows all students to be successful without modifying the text significantly.

Teacher Notes

TOPIC A: CREATING COLOURS: IDENTIFYING COLOURS

Learning Intentions:
Students will identify colours in a variety of natural and manufactured objects. Students will then arrange objects to recreate a rainbow.

Success Criteria:
- accurately use colour words like red, orange, yellow, green, blue and purple (or violet) to describe objects, either verbally or in written form.
- construct or illustrate or draw a rainbow with colours in the proper order (ROYGBP), using objects, paints or other art supplies.

Materials Needed:
- a copy of the *Identifying Colours* Worksheet 1 for each student
- pencils
- 6 plastic bags, each containing examples of food of one colour (red, orange, yellow, green, blue, purple)
- 6 containers, each containing coloured items (examples: toys or building blocks)
- 6 crayons, one each of the colours red, orange, yellow, green, blue, purple
- 6 markers, one each of the colours red, orange, yellow, green, blue, purple
- 6 construction paper cards or something similar, one each of the colours red, orange, yellow, green, blue, purple

Procedure:
1. Ask students to raise their hands if they have ever seen a rainbow. Tell the students a rainbow always begins with the colour red on the outside. Ask if the students can find anything in the classroom that has the colour red.

2. Lead the students in a similar discussion about each colour in order – orange, yellow, green, blue and purple. Purple is the last colour we see in a rainbow. Identify objects in the classroom for each colour.

3. Ask students if they can say the colours of the rainbow in order. Prompt them to start with red. If needed, model the order for them by saying "Red, orange, yellow, green, blue, purple." Lead the class with repetitions and then get the class to say it as a group.

4. Tell the students they can find these different colours in the foods that people eat. Prompt students to give examples of food for each colour (examples: red strawberries, orange oranges, yellow lemons or bananas, green lettuce or broccoli, blue blueberries and purple grapes or eggplants.

5. Tell the student they can find different colours in everyday objects too. Prompt students to give examples of toys or clothing or other manufactured objects for each colour (red, orange, yellow, green, blue, purple). If a colour is missing, find pictures that display the colours.

6. Read through the Identifying Colours Worksheet 1 with students. Students can then work independently on the Worksheet.

7. Divide students into groups of 6. Assign colours to each student in each group using the 6 plastic bags, the six containers, 6 crayons, 6 markers or 6 construction paper cards. (Be sure to give responsible students the more sensitive materials). Then tell students they must arrange their groups by colour in order to make a rainbow. Each group can either present to the class or report from their groups.

Differentiation:
Worksheet tasks can be shared in pairs or groups to accommodate different learners.

For slow learners, use a mnemonic such as ROYGBP or a song for the order of the names of the rainbow (some ideas on the website http://goo.gl/RJkQYG).

For enrichment, students could colour the Worksheet items and practice printing out the names of the foods on the Worksheet. As an extension, students could also arrange items within each colour (for example, from less blue to more blue).

Worksheet 1

Name: _____

Identifying Colours

Draw a line and match the colour with the right foods. You may use the colours more than once.

Red

cherries

carrot

Orange

broccoli

lemon

Yellow

lettuce

banana

Green

Blue

grapes

corn

tomato

Purple

peas

cherries

Teacher Notes

PRIMARY AND SECONDARY COLOURS

LEARNING INTENTION:
Students will learn that red, yellow, and blue are primary colours that can be used to create secondary colours such as orange, green, and purple.

SUCCESS CRITERIA:
- identify the primary colours that can be mixed to create secondary colours
- make predictions and investigate which primary colours can be mixed to make a secondary colour
- investigate how white light can be separated into colour
- record observations in drawings and through written descriptions
- make conclusions about colour creation

MATERIALS NEEDED:
- paper plates
- paint brushes
- red paint, yellow paint, blue paint, white paint
- glass prism(s)
- flashlight(s)
- a copy of *Primary and Secondary Colours* Worksheet 1 and 2 for each student
- a copy of *Experimenting With Colour* Worksheet 3 and 4 for each student
- pencils

PROCEDURE:
1. Lead students into a brainstorming activity/discussion about what the meaning is of primary and secondary. Once students come to understand that primary means first and secondary means second, tell students that basic colours we know can be put into a primary or secondary category. Ask students to name colours they think are the primary colours. (Primary colours are red, yellow, and blue). Explain to students that two primary colours can be mixed together to make a secondary colour. Students will investigate this idea by conducting an experiment with paint.

2. Read through the question, materials needed, and what to do sections on Worksheet 1. Students will make predictions, then conduct the investigation. They will record their observations and make conclusions about the mixing of primary colours to create secondary colours on Worksheet 2.

3. Students will continue to experiment with colour by adding white paint and black paint to the primary colours. They will record what they observe on Worksheet 3.

4. Divide students into pairs or small groups. Give each group a glass prism and a flashlight. Students will shine white light through the prism and record their observations on Worksheet 4.

DIFFERENTIATION:
Slower learners may benefit by working in a small group of three to complete the *Primary and Secondary Colours* activity. Each group member could be assigned a colour mixing task then share results with other members in the group in order to record all results needed to complete the Worksheet.

For enrichment, faster learners could paint a picture of a rainbow, based on their observations of colours.

Worksheet 1 Name: _____

Primary and Secondary Colours

Question: What primary colours could be mixed together to make a secondary colour?

Materials Needed

- 6 paper plates
- 6 paint brushes
- red paint, yellow paint, blue paint

What to do

1. Make predictions about the answer to the question. Record them on your Worksheet.

2. Put some red paint and some yellow paint on to a paper plate, use a paint brush to mix the colours together.

3. What secondary colour is made? Record your observations in the chart on your Worksheet.

4. Put some blue paint and some yellow paint on to a paper plate, use a paint brush to mix the colours together.

5. What secondary colour is made? Record your observations in the chart on your Worksheet.

6. Put some blue paint and some red paint on to a paper plate, use a paint brush to mix the colours together.

7. What secondary colour is made? Record your observations in the chart on your Worksheet.

Worksheet 2 Name: _____

Primary and Secondary Colours

Let's Predict

What will happen if red and yellow are mixed together?
_____

What will happen if blue and yellow are mixed together?
_____

What will happen if blue and red are mixed together?
_____

Let's Investigate

Mix the primary colours in the chart below. Colour the circle to show what secondary colour is made and label it.

Primary Colour Combinations			Secondary Colours
○ red	+	○ yellow	= ○ _____
○ blue	+	○ yellow	= ○ _____
○ blue	+	○ red	= ○ _____

| Worksheet 3 | Name: _____ |

Experimenting with Colour

Mix some white paint with each of the primary colours red, yellow, and blue. Describe the changes you saw for each colour.

This is what happened...

When I mixed red with white:

_____

When I mixed yellow with white:

_____

When I mixed blue with white:

_____

Mix some black paint with each of the primary colours red, yellow, and blue. Describe the changes you saw for each colour.

This is what happened...

When I mixed red with black:

_____

When I mixed yellow with black:

_____

When I mixed blue with black:

_____

Worksheet 4   Name: _____

Experimenting with Colour

We can find primary and secondary colours in a rainbow. When white light from the sun travels through water or glass, it bends and separates the white light into the colours we see in a rainbow. Let's investigate this idea!

You'll need:

- a glass prism
- a flashlight

What to do:

1. Shine the flashlight through the glass prism.
2. Observe and draw what you saw happen.

Let's Observe

This is a picture of what I saw.

[   ]

What colours did you see when the white light went through the prism?

_____

_____

Teacher Notes

APPLYING COLOURS

LEARNING INTENTIONS:

Students will distinguish between colours that are transparent and opaque. Students will apply paint to different surfaces to investigate how paint adheres to different materials. Students will extract a colour dye from one object and use it on another object.

SUCCESS CRITERIA:

- accurately identify transparent and opaque colours in paint or objects
- investigate how paint sticks to different surfaces, drawing conclusions from the experiments
- demonstrate and suitably explain how colour is transferred from one object to another using experience from experiments and their conclusions

MATERIALS NEEDED:

- copy of *Applying Colours* Worksheet 1, 2, 3 and 4 for each student
- coloured construction paper sheets, any colour
- coloured transparent filters or plastic lighting gels (certain plastic bags can be used as a substitute)
- paint brushes
- colour paint
- cup of water
- styrofoam plates or aluminum trays for paint palettes
- thick white paper for painting
- pieces of white cotton fabric large enough to be painted
- plastic containers
- beets cut in half

PROCEDURE:

1. Introduce students to the idea of things being transparent or opaque. Hold up a sheet of construction paper and ask *If I tried, could I look through this? If not, it means this sheet is opaque.* Hold up a transparent coloured filter and ask, "If I tried, could I look through this? If yes, it means this sheet it transparent." Use other items to lead students into using the two words fluently.

2. Introduce students to the phrases **brighter than** and **darker than**. Explain that with paint, we can make things brighter, darker, transparent or opaque.

3. Read through Applying Colours Worksheet 1 and 2 with the students. Demonstrate the procedure as needed. Prepare students by asking them for predictions of what will happen, and how they can test their predictions. Then monitor the work as students complete the worksheet.

4. Read through Applying Colours Worksheet 3 with the students. Demonstrate the procedure as needed. Prepare students by asking them for predictions of what will happen, and how they can test their predictions. Then monitor the work as students complete the worksheet.

5. Read through Applying Colours Worksheet 4 with the students. Demonstrate the procedure as needed. Prepare students by asking them for predictions of what will happen, and how they can test their predictions. Then monitor the work as students complete the worksheets.

DIFFERENTIATION:

Work through the entire procedure for each Worksheet as a class by performing the experiment for the students. They can record their predictions, observations and conclusions while remaining in their seats. For direct assistance, set up the procedures ahead of time and bring students or pairs to the scientific discovery station while the rest of the class completes seat work. Assign Worksheets to groups of students so that the work and procedures can be better managed together.

| Worksheet 1 | Name: _____ |

Applying Colours

We know that we can add white paint to coloured paint to get a different shade. Now let's investigate what happens when water is added to paint.

You'll need:

- a paint brush
- any colour of paint
- a cup of water
- 2 Styrofoam plates or aluminum trays
- a piece of white paper

What to do:

1. Pour some paint onto a Styrofoam plate.

2. Dip your paint brush into the paint on the plate, then paint 2 straight lines across your paper.

3. Dip your paint brush into the paint on the plate again. Paint over top of 1 of the lines again.

4. Pour some paint onto the other Styrofoam plate. Add a small amount of water and mix it together with your paint brush.

5. Using this new mixture, paint a straight line across your paper.

6. Make conclusions about what you have observed.

| Worksheet 2 | Name: _____ |

What I Observed

When I painted over one of the lines again, the colour became **(circle one)**:

lighter   darker   transparent   opaque   thicker   thinner

When I added a small amount of water and mixed it with the paint, the colour became **(circle one)**:

lighter   darker   transparent   opaque   thicker   thinner

What I Conclude:

Fill in the blanks.

More paint can make the colour appear _____ .

Mixing paint and water can make the colour appear _____ .

| Worksheet 3 | Name: _____ |

Applying Colours

We have had a lot of experience with painting on paper. Let's investigate how well paint sticks to other surfaces.

Question: Does paint stick to different surfaces like fabric or plastic?

You'll need:

- any colour of paint
- a paint brush
- a Styrofoam plate
- a piece of paper
- a piece of fabric
- a plastic container

What to do:

1. Make predictions about the answer to the question. Record them on your Worksheet.
2. Put some paint onto a paper plate.
3. Using your paint brush, paint a line across a piece of paper.
4. Now paint a line across a piece of fabric.
5. Now paint a line on the plastic container.
6. Make conclusions about what you observed.

What I Observed

Does paint stick to paper?	Yes	No
Does paint stick to fabric?	Yes	No
Does paint stick to plastic?	Yes	No

| Worksheet 4 | Name: _____ |

Applying Colours

Colour can be found in some vegetables. Vegetable dye is a natural way to add colour to fabric. Let's investigate this idea!

You'll need:

- a beet cut in half
- a piece of white cotton fabric

What to do:

1. Place the beet (cut side down) onto the white fabric. Continue to blot it on your piece of fabric so that it makes a design.

2. Record your observations by drawing what you created.

Let's Observe

This is a picture of the design I created on my piece of fabric.

Teacher Notes

TOPIC B: SEASONAL CHANGES: SEASONAL CHANGES

LEARNING INTENTION:

Students will learn about the four seasons, in terms of weather, clothing, activities, and how to prepare for each season.

SUCCESS CRITERIA:

- identify the four seasons of the year
- recognize the start and finish of each season
- recognize the types of weather in each season
- describe how to prepare for seasonal changes
- determine appropriate clothing needs for each season
- describe activities that are usually done in each season
- identify activities that can or cannot be done in other seasons
- make connections to their personal environment and daily lives

MATERIALS NEEDED:

- a copy of *A Look at the Four Seasons* Worksheet 1 for each student
- a copy of *What's the Weather?* Worksheets 2 and 3 for each student
- a copy of *Dressing for the Weather* Worksheets 4 and 5 for each student
- a copy of *Staying Active All Year Round!* Worksheets 6 and 7 for each student
- a copy of *When Does It Happen?* Worksheets 8 and 9 for each student
- a copy of *Just Where Did That Snowman Go?* Worksheet 10 for each student
- a copy of *A Seasonal Crossword* Worksheet 11 for each student
- chart paper, markers, pencils
- read aloud about the seasons in the year (*see suggestions in #1 of procedure section*)
- shoe boxes, plasticene, construction paper, scissors, glue (*optional materials*)

PROCEDURE:

*This lesson can be done as one long lesson, or be divided into four or five shorter lessons.

1. Introduce to students the four seasons in the year, ensuring their understanding that the seasons are cyclical. Discuss and display a visual of this yearly cycle on chart paper, indicating the months of the year that are associated with each season. At this point, it may be beneficial to do a read aloud about the seasons in order to give students a clear picture of what each season looks like. Suggested books are *What Makes the Seasons?* (Author: Megan Cash Montague), or *Caps, Hats, Socks, and Mittens* (Author: Louise Borden).

2. Ask students to refer back to the story or to recall from their own knowledge, as they explain the type of weather each season typically brings. Discussion should include such vocabulary such as precipitation type, temperature, windy, cloudy, sunny, etc. As a follow up, give students Worksheets 1, 2, and 3 to complete.

3. Divide students into pairs or into small groups. Give each group a piece of chart paper and markers. Ask students to discuss with their group members what people do to **prepare** for each season. Instruct them to divide their chart paper in four sections (one section per season), then draw and label the ideas they came up with. Some sample answers may be: Winter – get out snow shovels, sharpen skates; Spring – wash windows; Summer – buy sunscreen; Fall – rake leaves, put away patio furniture.

   *An option is to divide the class into four groups, then assign each group a season. Once groups are completed, they can present their findings to the large group. This promotes more discussion on how to prepare for seasonal changes.

4. Come together as a large group and discuss other ideas of how we know when seasons are changing (weather changes, wardrobe changes, changes in people's activities). Have students engage in a **knee to knee, eye to eye** activity where they turn and talk with a partner to brainstorm the types of weather each season brings, and the type of

clothing that is appropriate for each season. Come back as a large group, after pairs have discussed each season and share responses. Record student responses on chart paper, then post for students to reference as they complete Worksheets 4 and 5.

5. Come together as a large group and discuss the types of activities that people participate in during each season. Once again, have students engage in a **knee to knee, eye to eye** activity where they turn and talk with a partner to brainstorm the types of activities that people can do in each season. Come back as a large group, after pairs have discussed each season and share responses. Record student responses on chart paper, then post for students to use for reference as they complete Worksheets 6 and 7.

6. Come together as a large group. Refer back to the student responses on chart paper from the previous activity, and discuss which of the activities listed for each season, could actually be done in other seasons of the year. Ask students to think about what innovations allow for some of these activities to be done in other seasons, (e.g., arenas can make ice throughout all seasons, so skating and playing hockey can be done all year round). Ask students to think about the advantages or disadvantages of making this happen. Record student responses on chart paper. Then post for students to use as reference as they complete Worksheets 8 and 9.

7. Give students Worksheet 10 where they must decide on the order of events.

8. Give students the crossword puzzle on Worksheet 11 to complete.

*If students have access to technology such as iPads, there are some good learning opportunities to enhance students' awareness of seasonal changes, in terms of weather patterns, wardrobe changes, and typical seasonal activities. Explore Apps such as **The Four Seasons – An Earth Day Interactive Children's Story Book** or **Seasons and Weather! Learn with Poko**.

Differentiation:

Slower learners may benefit by working in a small group with teacher support to complete the written response component on Worksheets 4 and 5. Redirect these learners to vocabulary listed on chart paper of student responses on this topic. An additional accommodation would be to have these learners work with a peer to complete the crossword puzzle.

For enrichment, faster learners could choose their favorite season and create a diorama to depict some of the activities they can do, and clothing they would wear during that season. Dioramas can be created using plasticene or modeling clay.

A Look at the Four Seasons

In each picture, draw what the tree would look like in each season of the year.

Label the season that matches each picture.

_____ _____

_____ _____

Worksheet 2 Name: _____

What's the Weather?

Each season of the year brings different weather.

Draw pictures to show the kind of weather that may happen in each season.

Label your pictures using the weather words from the box below.

| rain | snow | wet | sunny | cloudy | thunder | humid |
| icy | hot | cool | foggy | warm | blizzard | lightning |

Weather in the Spring

Weather in the Summer

OTM-2170 ISBN: 978-1-4877-0198-7 © On The Mark Press

Worksheet 3 Name: _____

What's the Weather?

Continue drawing pictures to show the kind of weather that may happen in each season.

Label your pictures using the weather words from the box.

rain	snow	wet	sunny	cloudy	thunder	humid
icy	hot	cool	foggy	warm	blizzard	lightning

Weather in the Fall

Weather in the Winter

| Worksheet 4 | Name: _____ |

Dressing for the Weather

As the seasons change, so does the weather.

It is important to dress for the weather.

Describe the weather that each season brings.

Then, draw yourself dressed for the weather.

Label the clothing you are wearing.

In the spring the weather is...

_____
_____
_____
_____
_____

Dressing for spring weather

In the summer the weather is...

_____
_____
_____
_____
_____

Dressing for summer weather

Worksheet 5   Name: _____

Dressing for the Weather

Continue describing the type of weather and drawing yourself dressed for that weather.

Don't forget to label the clothing you are wearing.

In the fall the weather is...
_____
_____
_____
_____
_____

Dressing for fall weather

In the winter the weather is...
_____
_____
_____
_____
_____

Dressing for winter weather

Worksheet 6　　　Name: _____

Staying Active All Year Round

You have learned about the many different activities that people like to do during all seasons of the year.

What do *you* like to do in each season?

Draw and label yourself doing some of your favorite seasonal activities.

Spring

Summer

Worksheet 7   Name: _____

Staying Active All Year Round

Continue drawing and labelling yourself doing some of your favourite seasonal activities.

Fall

Winter

| Worksheet 8 | Name: _____ |

When Does it Happen?

Take a look back at the activities that you drew yourself doing in each of the seasons.

Make a list of the activities that you could do in **any** season.

1. _____
2. _____
3. _____
4. _____
5. _____
6. _____
7. _____
8. _____
9. _____
10. _____

Choose one of the activities from your list.

Explain why you can do this activity in any season.

_____
_____
_____
_____

| Worksheet 9 | Name: _____ |

When Does it Happen?

Take a look back at the activities that you drew yourself doing in each of the seasons.

Make a list of the activities that you can do only in **certain** seasons.

1. _____
2. _____
3. _____
4. _____
5. _____
6. _____
7. _____
8. _____
9. _____
10. _____

Choose one of the activities from your list.

Explain why you can only do this activity in a certain season.

_____
_____
_____
_____

Just Where Did That Snowman Go?

Look at the pictures below. They are out of order.

Number the pictures in the correct order to tell a story.

Why do you think the snowman melted? _____
_____
_____

Worksheet 11   Name: _____

A Seasonal Crossword

Read the clues and use words from the Word Box to help you fill in the answers to this crossword puzzle.

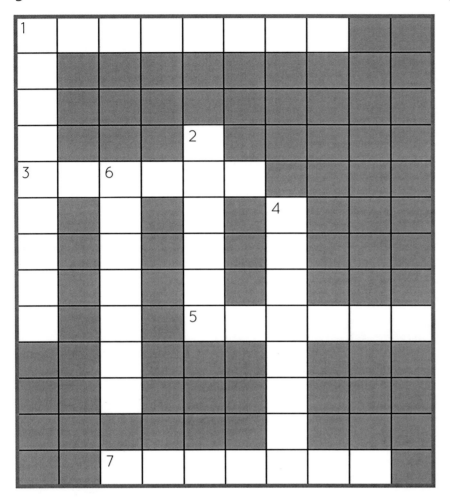

Word Box

mittens
umbrella
spring
leaves
snowstorm
summer
sledding
January

ACROSS
1. This activity is done in the winter.
3. We can swim at the beach during this season.
5. We plant flowers in this season.
7. This is the first month of the year.

DOWN
1. This storm can happen in the winter.
2. We rake these in the fall.
4. We use this to stay dry from the rain.
6. We wear these in the winter to stay warm.

Teacher Notes

ABORIGINAL ACTIVITIES

LEARNING INTENTION:
Students will learn about the activities of Aboriginals throughout seasonal changes.

SUCCESS CRITERIA:
- describe the seasonal activities of a First Nations people of long ago
- compare a seasonal activity of a First Nations people of long ago to one of your own
- conduct an interview with an Aboriginal Elder to learn about seasonal activities of today
- compare seasonal activities of a First Nations people of long ago to seasonal activities of a First Nations people of today

MATERIALS NEEDED:
- a copy of *Circle of Seasons* Worksheet 1 for each student
- a copy of *Comparing Activities* Worksheet 2 for each student
- a copy of *Interview with an Elder* Worksheets 3 and 4 for each student
- a copy of *Comparing Activities – Part Two* Worksheet 5 for each student
- a read aloud (*see #1 in procedure section*)
- chart paper, markers
- clipboards (one per student)
- pencils

PROCEDURE:
*This lesson can be done as one long lesson, or be divided into two shorter lessons.

1. Discuss with students the meaning of 'First Nations people' or 'Aboriginals', to ensure students' understanding of who they are. Accessing photographs on-line would aid students in understanding their lifestyle in terms of their seasonal activities.

2. A suggested read aloud at this point, is *Moonstick* by Eve Bunting. On chart paper, draw a large circle and divide it into 4 quadrants labeled Spring, Summer, Fall, Winter. While reading aloud, pause after the description of each moon. Ask students to use the illustrations and descriptions to determine what season of the year it is. Also, ask students to describe the activities that the Aboriginal people do during that season. Record students' responses on the chart paper inside the appropriate quadrant.

3. Give students Worksheet 1 to complete. Students can refer to the recorded responses on the chart paper to help them draw their pictures.

4. Give students Worksheet 2 to complete. Students can refer back to Worksheets 6 and 7 from the previous lesson for the responses they completed about the seasonal activities that they participate in.

5. **Invite an Aboriginal Elder** from a local community to come in to speak to the students about their traditional seasonal activities. Give each student a clipboard, and Worksheets 3 and 4 to complete as they conduct the interview. Students may write the responses in point form or draw pictures of the seasonal activities that are being described to them.

6. Give students Worksheet 5 to complete. Essentially, they will compare seasonal activities of a First Nations people from long ago to ones of today.

DIFFERENTIATION:
Slower learners may benefit by working in a small group with teacher support to complete the written descriptions of the seasonal activities on Worksheets 2 and 5.

For enrichment, faster learners could prepare three additional questions to ask the Elder.

Worksheet 1   Name: _____

Circle of Seasons

In the story *Moonstick*, the young boy watches the changing seasons and learns about the activities that his people do. He learns that changes come and will come again, like a circle of life.

Use the circle chart to draw the activities that his people did during each of the seasons.

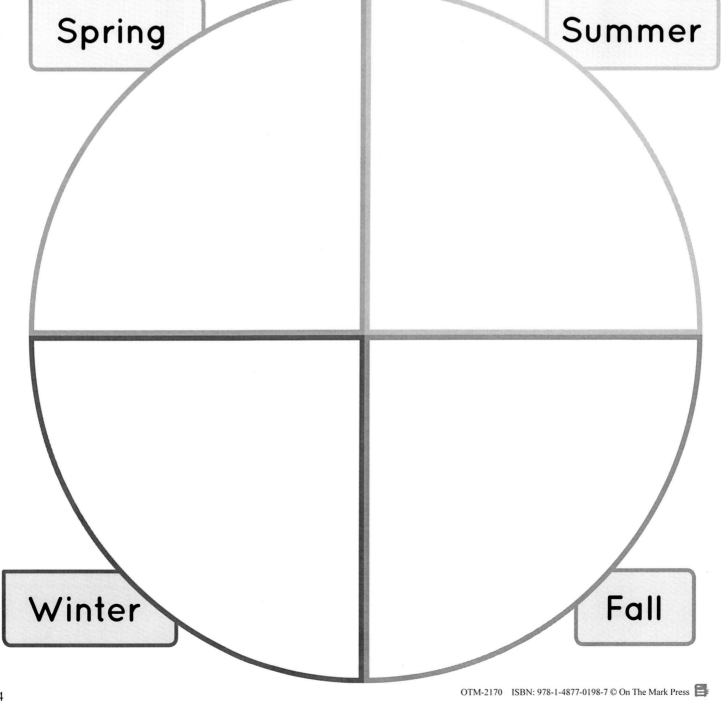

| Worksheet 2 | Name: _____ |

Comparing Activities

Choose one of the seasons in the circle on Worksheet 1.

Describe the activities that the boy's people did during that season.

Season: _____

_____
_____
_____
_____
_____
_____
_____

Compare these activities to the ones that you do in that season today.

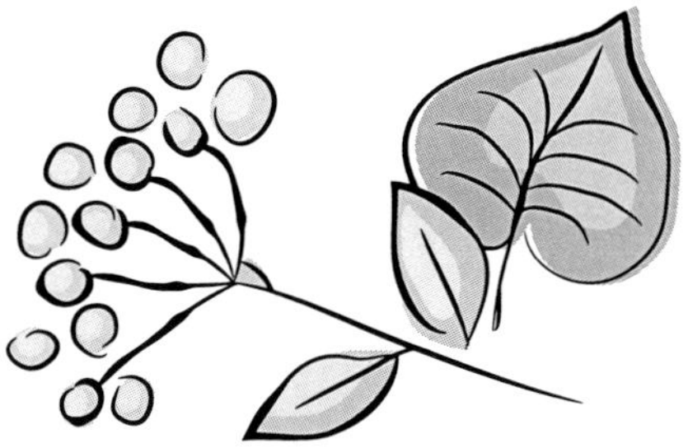

What is the **same**?
What is **different**?

_____
_____
_____
_____
_____
_____

Worksheet 3   Name: _____

Interview with an Elder

Interview an Aboriginal Elder. Use the questions below as a guide when asking the Elder about different seasonal activities that his/her people do today.

1. What are some activities that your people do in the spring time?

2. What are some activities that your people do in the summer time?

Worksheet 4   Name: _____

Use the questions below as a guide when asking the Elder about different seasonal activities that his/her people do today.

3. What are some activities that your people do in the fall time?

4. What are some activities that your people do in the winter time?

| Worksheet 5 | Name: _____ |

Comparing Activities – Part Two

Choose another one of the seasons in the circle on Worksheet 1.

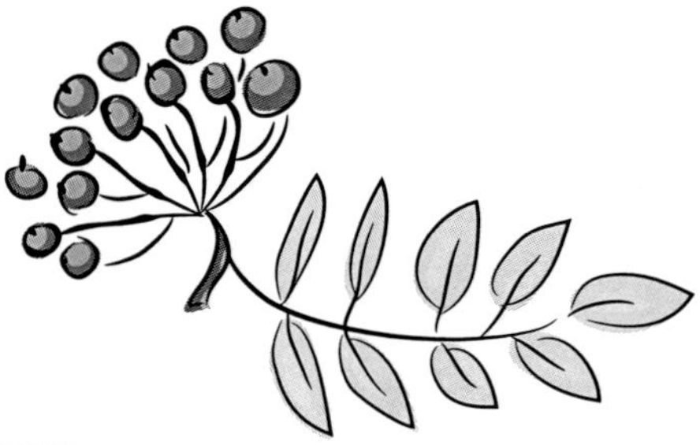

Describe the activities that the boy's people did during that season.

Season: _____

_____

_____

_____

_____

_____

Compare the activities of the boy's people to the activities of this same season that you learned about from your interview with an Aboriginal Elder of today.

What is the **same**?
What is **different**?

_____

_____

_____

_____

_____

Teacher Notes

THE HEAT OF THE SUN

LEARNING INTENTION:
Students will learn about heat from the sun that happens throughout the day and the seasons.

SUCCESS CRITERIA:
- identify the benefits of the sun
- determine temperature readings and make connections to appropriate clothing
- recognize differences in the sun's heat throughout the day
- identify ways to keep comfortable at different temperatures
- make connections to their daily lives

MATERIALS NEEDED:
- a copy of *Help From the Sun* Worksheet 1 for each student
- a copy of *Feel the Heat* Worksheets 2 and 3 for each student
- a copy of *Taking a Reading* Worksheets 4 and 5 for each student
- a copy of *What to Wear?* Worksheet 6 for each student
- a copy of *What's the Temperature?* Worksheets 7, 8, 9, 10 for each student
- a copy of *Get Comfortable!* Worksheet 11 for each student
- a warm sunny day, **students are to wear sunscreen, hat, and sunglasses**
- a large thermometer (teaching thermometer)
- outdoor thermometers (one for each student or pair of students)
- chart paper, markers, rulers
- pencils

PROCEDURE:
*This lesson can be done as one long lesson, or be divided into four or five shorter lessons.

1. Discuss with students the idea that the sun is the Earth's main source of heat and light. Divide students into pairs and give them Worksheet 1. They will work with a partner to **think, pair,** **share** an answer to the question. A next step would be to encourage students to share their responses to the large group. This would lead to some rich discussion about the impact that the sun has on our lives.

2. On a warm sunny day, take students out for a walk. (Ensure they wear sunscreen, hat, and sunglasses). Give them Worksheets 2 and 3, a clipboard, and a pencil. It may be beneficial to discuss what **'describe'** and **'predict'** mean before starting this activity. Inform students that they are to make some observations about how the sun feels on their skin. Once the observation is completed, an option would be to return to the classroom and discuss their findings.

3. Using a large (teaching) thermometer, teach students how to read the temperature on a thermometer. Ensure students are able to relate a certain temperature to a type of weather, and determine what clothing to wear. Give students Worksheets 4, 5, and 6 to complete.

4. Brainstorm with the students a list of words to describe temperature and the outside air. Some examples are hot, warm, humid, mild, cool, cold, chilly, freezing, etc. They will need to refer to this vocabulary each day as they take temperature readings on Worksheets 7, 8, 9, and 10. *Daily times for readings can be adjusted, depending on the schedule of your school day.

   Extended teaching option:
   - revisit Activity #4 in the procedure section, using Worksheets 7, 8, 9, and 10 at a **different time of year** in order to compare the seasonal results of the sun's heat.

5. Have a discussion with students about the features of buildings that keep people sheltered and comfortable throughout daily and seasonal changes. This discussion may lead to other ideas about our **creature comforts** that keep us comfortable in different temperatures. Some sample answers may be fireplaces, air conditioners, fans, windows, pools, access to beaches, hot or cold beverages. Give students Worksheet 11 to complete individually or with a partner.

| Teacher Notes |

DIFFERENTIATION:

Slower learners may benefit by:

- completing only pictures in order to provide answers on Worksheet 1
- working with a peer to complete the observational notes on Worksheets 2 and 3
- working in a small group with teacher direction to ensure they obtain accurate temperature readings for Worksheets 4, 7, 8, and 9
- completing Worksheet 10 together as a small group, with teacher direction, to reduce the amount of written output required from these learners

For enrichment, faster learners could choose one day of the week from Activity #4 in the procedure section, and make a bar graph to represent the temperature differences throughout the day. An option would be to have these learners create a bar graph for each of the 5 days that they took temperature readings.

Worksheet 1 Name: _____

Help From the Sun

Think, Pair, Share

With a partner, do some thinking and sharing of ideas about the question below.

Use pictures and words when recording your ideas in the chart.

"How does the sun help us?"

My Thinking

_____
_____
_____
_____
_____
_____

My Partner's Thinking

_____
_____
_____
_____
_____
_____

Worksheet 2   Name: _____

Feel the Heat

Take a walk in the sun on a warm sunny day.

Observe how the sun feels on your skin.

Come on, let's go!

Describe how your skin feels after a few minutes in the sun.

_____
_____
_____
_____

When you stay in the sun for a long time on a **hot** day, what happens to your body?

_____
_____
_____
_____
_____

| Worksheet 3 | Name: _____ |

Let's Predict

Predict how you will feel if you go into the shade.

_____
_____
_____

Let's Observe

Now find a place in the shade.

Observe how your skin feels.

Describe how your skin feels after a few minutes in the shade.

_____
_____
_____
_____

Let's Conclude

What can you conclude about the heat of the sun?

_____
_____
_____
_____

Worksheet 4   Name: _____

Taking a Reading

A thermometer measures the temperature. It helps us to know how hot, how warm, or how cold things are.

We can use a thermometer to measure the temperature outside.

Let's take some readings! Record the temperature shown on each thermometer, and circle Celcius (°C) or Fahrenheit (°F).

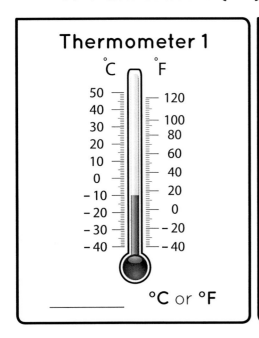

_____ °C or °F

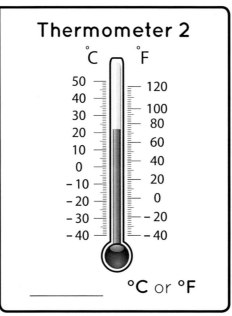

_____ °C or °F

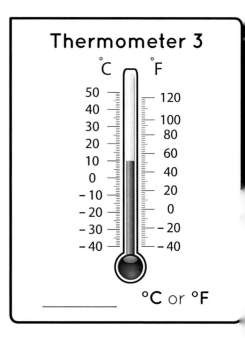

_____ °C or °F

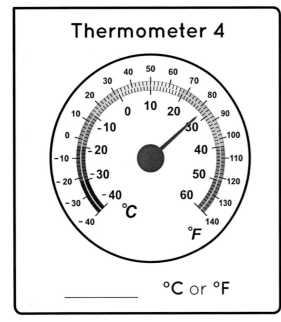

_____ °C or °F

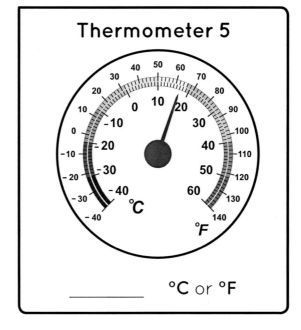

_____ °C or °F

Worksheet 5   Name: _____

Use **warmer than** or **cooler than** to compare the temperatures on the thermometers on Worksheet 4.

Thermometer 3 is _____ **Thermometer 1.**

Thermometer 3 is _____ **Thermometer 2.**

Thermometer 4 is _____ **Thermometer 5.**

Thermometer 2 is _____ **Thermometer 5.**

Which thermometer on Worksheet 4 could be used to describe the temperature in each picture below?

Thermometer ___   Thermometer ___   Thermometer ___

| Worksheet 6 | Name: _____ |

What to Wear?

Read the temperature on the thermometers.

Make a list of the clothing items that you should wear.

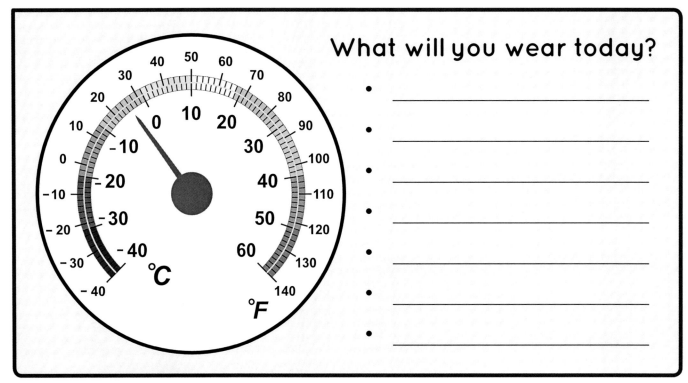

What will you wear today?
- _____
- _____
- _____
- _____
- _____
- _____
- _____

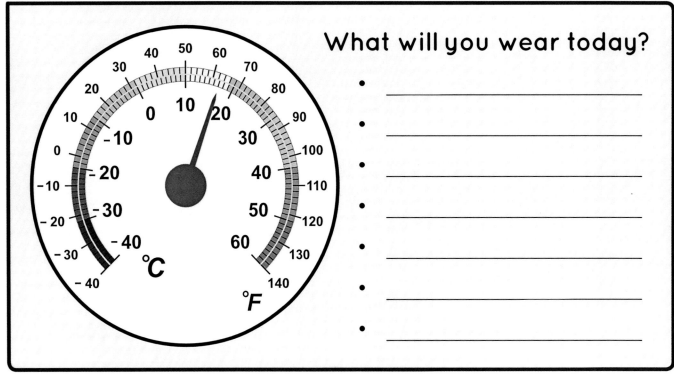

What will you wear today?
- _____
- _____
- _____
- _____
- _____
- _____
- _____

What's the Temperature?

Now that you know how to read a thermometer, let's take some temperature readings during the day to learn more about the sun's heat.

Measure and record the outdoor temperature at different times of the day, for 5 days.

Each time you do, record how the outside air feels.

Let's Observe

Monday

Time	Temperature Reading	Describe the outside air
9:00am		
11:30am		
2:00pm		

Worksheet 8   Name: _____

Tuesday

Time	Temperature Reading	Describe the outside air
9:00am		_____ _____ _____
11:30am		_____ _____ _____
2:00pm		_____ _____ _____

Wednesday

Time	Temperature Reading	Describe the outside air
9:00am		_____ _____ _____
11:30am		_____ _____ _____
2:00pm		_____ _____ _____

Worksheet 9　　　　　　　　Name: _____

Thursday

Time	Temperature Reading	Describe the outside air
9:00am		_____ _____ _____
11:30am		_____ _____ _____
2:00pm		_____ _____ _____

Friday

Time	Temperature Reading	Describe the outside air
9:00am		_____ _____ _____
11:30am		_____ _____ _____
2:00pm		_____ _____ _____

| Worksheet 10 | Name: _____ |

Let's Conclude

Are all the morning temperatures (9:00am) similar? Explain. _____
_____
_____

When was the warmest temperature of a day? Explain why. _____
_____
_____

Why is it important to know when the warmest temperature of a day usually happens?
_____
_____
_____

When was the coolest temperature of a day? Explain why. _____
_____
_____

Why is it important to know when the coolest temperature of a day usually happens?
_____
_____
_____

Worksheet 11   Name: _____

Get Comfortable!

You have learned that as the seasons change, the temperature changes.

We wear different types of clothing in order to be comfortable in different types of weather.

What other things would help you to stay comfortable in different temperatures?

Draw and label your ideas.

In the Heat

In the Cold

Teacher Notes

ANIMAL ADAPTATIONS

LEARNING INTENTION:

Students will learn about the changes of animal appearance, location, and activity throughout seasonal changes.

SUCCESS CRITERIA:

- recognize the differences in some animals' appearance according to the season
- determine the reasons for differences in some animals' appearance in the seasons
- identify the migratory path of the Monarch butterfly
- determine reasons for migration
- name some hibernating animals
- identify the signals that hibernating animals use to recognize seasonal change

MATERIALS NEEDED:

- a copy of *Adapting to Seasons* Worksheet 1 for each student
- a copy of *A New Coat* Worksheet 2 for each student
- a copy of *Migration* Worksheet 3 for each student
- a copy of *Time to Hibernate* Worksheets 4, 5, and 6 for each student
- a copy of *Adapt an Animal* Worksheet 7 for each student
- a read aloud about migration *(see suggestion in #3 of procedure section)*
- a read aloud about hibernation *(see suggestion in #5 of procedure section)*
- chart paper, markers
- modelling clay
- pencils

PROCEDURE:

*This lesson can be done as one long lesson, or be divided into three or four shorter lessons.

1. Give students Worksheet 1. Read it through with them to ensure their understanding of migration and hibernation. Brainstorm some animals that experience changes in appearance, location, or activity depending on seasonal differences. Record students' responses on chart paper, under the three headings of **appearance, location,** and **activity**.

   (Teacher Note: Examples of animals that change fur colour in winter months are an Arctic fox, snowshoe hare, Arctic hare, a ptarmigan, caribou. Examples of animals that change location are Canada geese, Monarch butterfly, whales, caribou, many types of birds. Examples of animals that hibernate are bears, bats, hedgehogs, snakes, turtles, squirrels, chipmunks.)

2. Give students Worksheet 2 to complete. After talking with a partner about the reasons for the change in the snowshoe hare's coat of fur, come back together as a group, and have a large group discussion about the reasons. Record students' responses on the previously used chart paper under the section of **appearance**. Choose a few more animals from that list and discuss what changes in appearance help them to survive seasonal changes.

3. Discuss with students the meaning of migration. At this point, it may be beneficial to do a read aloud about migration in order to further their understanding. A suggested book is *Monarch Butterfly* (Author: Gail Gibbons). After reading, discuss how a Monarch butterfly spends it day, what *signals* it uses to recognize that the season is changing, and the places it migrates to (in Fall, and in the Spring). Give students Worksheet 3 to complete.

4. Have a large group discussion about the reasons why Monarch butterflies migrate. Record students' responses on the previously used chart paper under the section of **location**. Choose a few more animals from that list and discuss **reasons** why a change in their location help them to survive seasonal changes.

5. Discuss with students the meaning of hibernation. At this point, it may be beneficial to do a read aloud about hibernation in order to further their understanding. A suggested book is *Time to Sleep* (Author: Denise Flemming). After reading, discuss what *signals* the animals in the story used to recognize that the season was changing and it was time to hibernate. Also, discuss what some animals *do to prepare* for their hibernation. Record students' responses on the previously used chart paper under the section of **activity (hibernation – signals, preparation)**. Give students Worksheets 4, 5, and 6 to complete.

6. Give students Worksheet 7. They will create their own animal depicted in its habitat, then indicate its adaptations that help it survive in its habitat. A follow up option is to do a **share fair**, displaying students' creations around the classroom, then allowing students to roam the room to see what "animals" their classmates created.

DIFFERENTIATION:

Slower learners may benefit by working in a small group with teacher direction to have a discussion about the reasons for the Monarch butterflies migration, prior to completing the written component on Worksheet 3. This same group may benefit from revisiting the story *Time to Sleep*, while they complete Worksheet 4.

For enrichment, faster learners could use modeling clay to create a three dimensional figurine of their animal that they created on Worksheet 7.

Adapting to Seasons

Some animals change the way they look, where they live, or their activities depending on which season of the year it is.

These changes help certain animals survive the changes in weather throughout the seasons.

An Arctic fox's **fur will change** from brown to white as the season of fall turns to winter.

Canada geese change location as the seasons change. In the middle part of fall, they begin to **migrate** south to a warmer place to spend the winter months. They return to Canada in the spring as the weather gets warmer.

 A black bear is an example of an animal that **hibernates** during winter. During hibernation a black bear's heart rate may drop from 50 to 8 beats per minute, as they sleep. A black bear can go as long as 100 days without eating or drinking, while in hibernation.

A New Coat

Below is a picture of a snowshoe hare in its summer coat of fur, and a picture of it in its winter coat of fur.

Describe how the snowshoe hare looks in the summer.

_____
_____
_____

Describe how the snowshoe hare looks in the winter.

_____
_____
_____

With a partner, talk about some reasons why the snowshoe hare's coat of fur changes as the seasons change.

Worksheet 3   Name: _____

Migration

Imagine it is fall and the Monarch butterflies are beginning their migration. Draw arrows to show the direction of the butterflies' migratory path.

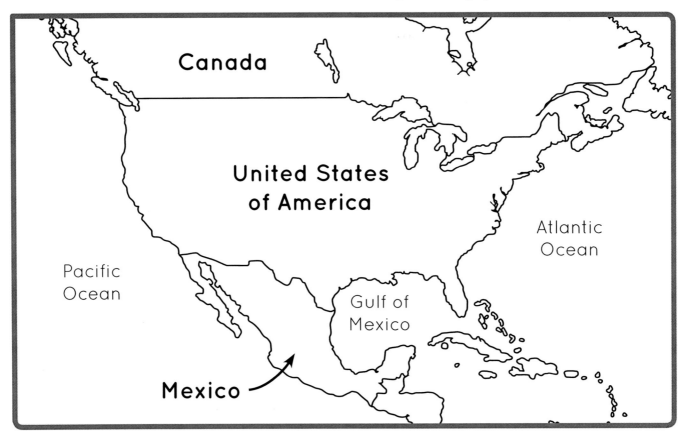

Where do the Monarch butterflies migrate to in the winter? _____
_____

Give some reasons why the Monarch butterflies migrate. _____
_____
_____
_____

Worksheet 4   Name: _____

Time to Hibernate

The animals in the story **Time to Sleep** watched for signs in nature to tell them that the season was changing from fall to winter.

What sign did each animal use to tell it was hibernation time? Use the ideas in the Word Box below to help you with your answers.

leaves change colour	days get shorter	leaves fall off the trees
frost on the grass	geese migrating	smell winter in the air

_____
_____
_____
_____

_____
_____
_____
_____

_____
_____
_____
_____

_____
_____
_____
_____

_____
_____
_____
_____

_____
_____
_____
_____

| Worksheet 5 | Name: _____ |

Time to Hibernate

Draw and describe where each of the animals in the story **Time to Sleep**, hibernate for the winter.

Bear sleeps _____

Snail sleeps _____

Skunk sleeps _____

Worksheet 6　　　　Name: _____

Draw and describe where each of the animals in the story **Time to Sleep** hibernate for the winter.

Turtle sleeps _____

Woodchuck sleeps _____

Ladybug sleeps _____

Worksheet 7

Name: _____

Adapt an Animal

Show your creativity! Create your own animal.

Be sure to:
- give your animal a name
- draw your animal in its habitat
- label the parts of your animal that help it to survive in its environment

Describe your animal's habitat.

_____
_____
_____

Explain why your animal is suited to its environment.

_____
_____
_____

Teacher Notes

PLANT ADAPTATIONS

LEARNING INTENTION:
Students will learn about the differences in the appearance and behaviour of plants throughout seasonal changes.

SUCCESS CRITERIA:
- describe the changes in the appearance of some plants throughout the four seasons
- describe the changes in the behaviour of some plants during the day and at night
- choose a plant, then depict the appearance of it in the current season
- imagine the plant or tree in a different season, depict its appearance

MATERIALS NEEDED:
- a copy of *A Change of Plants!* Worksheets 1 and 2 for each student
- a copy of *A Dandelion Throughout the Seasons* Worksheet 3 for each student
- a copy of *A Look at Plant Life* Worksheets 4 and 5 for each student
- a copy of *What Is It Like To Be a Tree?* Worksheet 6 for each student
- a read aloud (see #2 in procedure section)
- clipboards, pencils (for each student)
- paint, paint brushes, white art paper (optional materials)

PROCEDURE:
*This lesson can be done as one long lesson, or be divided into two shorter lessons.

1. Give students Worksheets 1 and 2. Read through the information with them to ensure their understanding of how some plants adapt to seasonal changes, and daily changes.

2. As a further resource of information, a suggested read aloud is *A Dandelion's Life* by John Himmelman. After reading, engage students in a discussion of how the dandelion plant adapts to seasonal and to daily change. An option is to have students take part in a **knee to knee, eye to eye** activity where they turn and talk about this topic with a partner. Come back as a large group to share responses. Some guiding questions are:
   - How does the dandelion plant change throughout the seasons of the year?
   - What does it look like in each season?
   - What does the dandelion do on an early summer evening?
   - Why do you think the dandelion closes its flower head?

3. Give students Worksheet 3 to complete.

4. Give students Worksheets 4 and 5, and a clipboard and pencil. Take students on a nature walk outside in the school yard or nearby area. Instruct them to find a tree or a plant, and draw a detailed picture of it on Worksheet 4. Their drawing should be representative of the season of the year you are in. Next, they will imagine what their tree or plant would look like in a different season, and draw it on Worksheet 5.

5. Divide students into pairs. Give them Worksheet 6 to complete. They will use creative movement to dramatize how a tree would feel in each situation listed.

DIFFERENTIATION:
Slower learners may benefit by working in a small group with teacher support to complete the written descriptions of the changes in the dandelion on Worksheet 3. It may be beneficial to have this group reference the *Life of a Dandelion* book as they complete the descriptions.

For enrichment, faster learners could paint a picture of the tree or plant they chose to draw. This art activity could also include a painting of their tree or plant in a different season.

Worksheet 1

Name: _____

A Change of Plants!

Plants, like animals, need to adapt to their environments. They do this in different ways.

Some plants lose their leaves in the winter.

Trees that do this are called **deciduous**. Deciduous trees lose their leaves to better survive winter weather.

They are able to conserve water this way, and this helps them to grow new leaves in the spring.

A **maple tree** is an example of a **deciduous** tree.

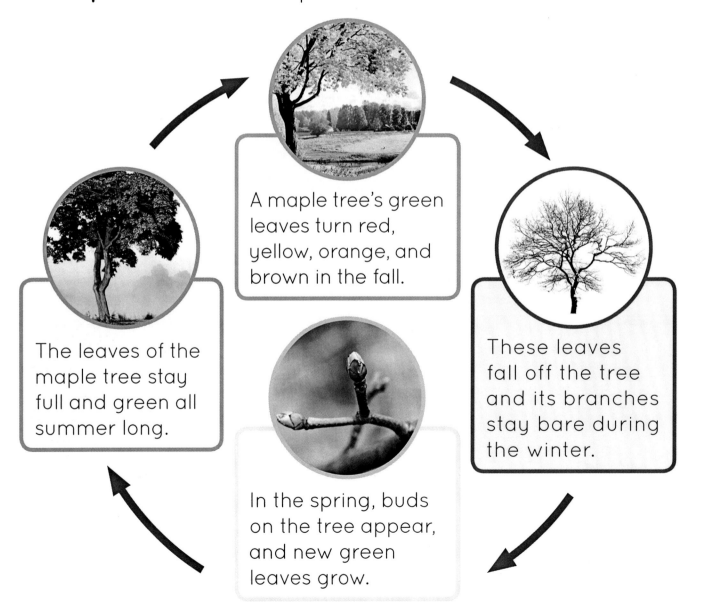

A maple tree's green leaves turn red, yellow, orange, and brown in the fall.

These leaves fall off the tree and its branches stay bare during the winter.

In the spring, buds on the tree appear, and new green leaves grow.

The leaves of the maple tree stay full and green all summer long.

Many bushes also lose their leaves when the frost comes.

These plants go into a winter sleep, called dormancy.

Plants which are dormant during the winter need the coldness so that they can grow again when spring comes.

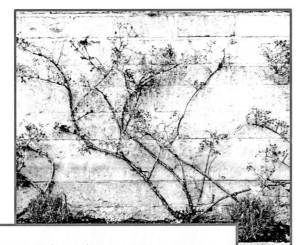

A rose bush goes dormant for the winter.

A rose bush blooms again in the spring.

Some plants adapt to their environment daily.

They do this by opening up in the daylight and closing up at night.

Some flowering plants do this so that they can protect their nectar or protect themselves from the cool temperature of the night.

A crocus plant opens its leaves in the light of day.

A crocus plant closes its leaves at night time.

A Dandelion Throughout the Seasons

Describe how the dandelion plant changes throughout the seasons of the year.

Draw diagrams to show its changes.

In the **spring**, a dandelion _____

In the **summer**, a dandelion _____

In the **fall**, a dandelion _____

In the **winter**, a dandelion _____

Worksheet 4   Name: _____

A Look at Plant Life

Draw a picture of a tree or a plant that is in your school yard.

Be sure to add details to your drawing.

Some things to think about as you draw are:

Are there leaves or flowers on it?

What colour are the leaves?

Does it have bare branches?

The season is _____

| Worksheet 5 | Name: _____ |

A Look at Plant Life

Draw a picture of your tree or plant in a **different season** of the year.

Be sure to add details to your drawing.

Imagine as you draw:

> What season is it?
>
> Is there a change in the leaves?
>
> Is there a change in the flowers?

The season is _____

Worksheet 6   Name: _____

What Is It Like To Be a Tree?

Working with a partner, use small and large movements to move like a tree for each situation in the list below.

- a tree in a soft spring breeze
- a tree in a strong windstorm in the fall
- a tree in the pouring rain
- a tree in a forest fire in the hot summer
- a tree without leaves in the winter
- a tree with a squirrel running up it
- a tree with a bird sitting on its branch
- a tree with a person climbing up it
- a tree that is being cut down

From the list above, choose a situation that you created movement for.

Draw what that tree would look like in nature in that situation.

Draw yourself creating the same movement of the tree.

Teacher Notes

TOPIC C: BUILDING THINGS: CONSTRUCTING A BIRDFEEDER

LEARNING INTENTION:

Students will learn using problem solving skills and knowledge acquired from previous investigations to design and build a birdfeeder.

SUCCESS CRITERIA:

- identify recyclable and reusable materials
- design and make a plan to construct a birdfeeder
- make and record observations of the final product by comparing and sharing similarities, differences, and benefits of the design
- make connections about action plans to reduce waste in the environment

MATERIALS NEEDED:

*Send a note home to parents explaining that their children are going to be building an item made from cleaned, recycled materials. A suggestive list of materials to have students bring in are: egg cartons, Styrofoam trays or plates, plastic and paper bags, boxes, plastic bottles and milk jugs, newspapers, wrapping paper, cardboard tubes, margarine tubs or other plastic containers

- other materials needed are glue, string, fasteners, paint, aluminum foil, markers, scissors, and boxes for sorting items brought from home
- a copy of *Constructing A Birdfeeder* Worksheet 1 and 2 for each student
- pencils

PROCEDURE:

1. The collection of recycled items from home is an opportunity for students to learn about recycling and reusing. Display the items for the students to examine.
   Ask them:

   > Where did these objects come from?
   > Why did you not just throw them in the garbage?
   > Why is it important to reuse and recycle?
   > How does this help the environment?

2. Explain to students that they are going to plan, design, and construct a birdfeeder made out of the recyclable materials. Then, they will compare their final product with a partner and make some observations.

3. Give each student a copy of the *Constructing A Birdfeeder* Worksheet 1 and 2 to complete, as they begin to construct their birdfeeder.

4. Once all birdfeeders are completed, invite students to present their birdfeeders to the class and share three things about their feeders.

DIFFERENTIATION:

Slower learners may benefit by listing only one thing that is the same and one thing that is different from their partner's birdfeeder, and provide only one thing to share with the class about their birdfeeders.

For enrichment, faster learners could build an additional object of their choosing, using items from the recycled materials collection.

Worksheet 1   Name: _____

Constructing a Birdfeeder

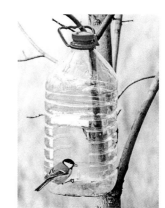

You know a lot about materials. Use what you know to design and construct a birdfeeder.

You'll need:

The materials I will use are:

_____   _____

_____   _____

_____   _____

My Plan and Design:

My plan for making the birdfeeder is:

1. _____
2. _____
3. _____
4. _____
5. _____

Draw a design of what your birdfeeder will look like:

Now gather your materials, and carry out your plan!

| Worksheet 2 | Name: _____ |

Constructing A Birdfeeder

Working with a partner, **compare** your final product.

In the chart below, list **two things that are the same** and **two things that are different** about your birdfeeders.

Things That Are The Same	Things That Are Different
1. _____ _____	1. _____ _____
2. _____ _____	2. _____ _____

Three things I would like to share with the class about my birdfeeder are:

1. _____
   _____

2. _____
   _____

3. _____
   _____

What could be done with your birdfeeder when it is no longer being used?

_____

_____

Teacher Notes

BUILDING A POP-UP CREATURE CARD AND BUILDING A TOY ANIMAL

LEARNING INTENTION:

Students will construct and decorate toys, recognizing how items are developed for certain purposes.

SUCCESS CRITERIA:

- select appropriate materials for the creation of a card or toy
- identify component parts and describe the purpose of those parts
- make conclusions about methods and materials involved in the development of their projects

MATERIALS NEEDED:

- copy of *Building a Pop-Up Creature Card* Worksheet 1, 2, and 3 for each student
- copy of *Building a Toy Animal* Worksheet 4 and 5 for each student
- copy of
- masking tape, scotch tape, glue sticks or white glue
- scissors
- pencil crayons, crayons, markers, pencils,
- white sheets of paper, coloured sheets of paper
- optional decoration items like small beads, glitter, ribbon
- toilet paper or paper towel tubes

PROCEDURE:

Ideas from websites:

- http://goo.gl/grrvUn
- http://goo.gl/5c35mP

Ideas from youtube:

- https://goo.gl/HtsYs6
- https://goo.gl/VoQVBD
- https://goo.gl/Jh6rHx
- https://goo.gl/HksmQw
- https://goo.gl/vzjR6t

1. Lead students in a discussion about faces. What kinds of things do faces have? Do faces have mouths? Do faces have eyes? Do faces have noses? Tell students they are going to make a card and inside that card there will be a face. They will have to figure out if it is a frog, a bird, some kind of monster creature or something else. (This lesson can also be used in conjunction with special days as cards to give to family members.)

2. Read through the question, materials and What I Do section of the Building a Pop-up Creature Card Worksheet with the class. Use prepared examples with each step. Ask students what kind of face or creature they think is in your example card. Ask each student why they think it would be that kind of creature. As a class, read through the What I Observe section of the Worksheet to prepare students to think about their answers after their work.

3. Students will build their creature cards and decorate them independently. When they are finished make sure they answer the questions on the Worksheet.

4. Tell the children that the creature card is an object that can lie flat but some objects will stand up. When you build an object to stand on its own, you can call it a model or a figure (or figurine). Explain to the children that in the next project they will make a toy animal that can stand on its own, a toy frog. Lead students in a discussion about the features of a frog. For example a frog has green skin. A frog has two eyes. A frog has a tongue, a frog has two legs and two arms.

5. Read through the question, materials and What I Do section of the Building a Toy Animal Worksheet with the class. Use prepared examples with each step. Ask students what kind of face or creature they think is in your example toy figure. Ask each student why they think it would be that kind of creature. As a class, read through the What I Observe section of the Worksheet to prepare students to think about their answers after their work. Lead students in a discussion on how they play with toy figures and things they might do.

> Teacher Notes

6. Students will build their toy frog figure and decorate them independently. When they are finished be sure they answer the questions on the worksheet.

DIFFERENTIATION:

Before the class, prepare some sheets of paper with the cuts already made to accommodate the slow learners or students who haven't yet developed skill or dexterity with scissors. Cut out some shapes from paper – triangles, circles, rectangles, others -

For enrichment, learners can make their pop-up card with pull tabs. Use these videos as guides for students:

https://goo.gl/qeWP3d

and

https://goo.gl/nxPEJQ

As an extension, learners can make a different animal figure besides the frog by using a different colour for the body, adding ears, or cutting legs out on the bottom of the roll. Some examples could include an octopus or a bear.

Worksheet 1 Name: _____

Building a Pop-Up Creature Card

What will your pop-up creature look like? Will it be a frog, a bird, a monster or something else?

You'll need:

- masking tape, scotch tape, glue sticks or white glue
- scissors
- pencil crayons, crayons, markers, pencils,
- white sheets of paper, coloured sheets of paper
- decorating items

What to do:

1. Fold one piece of paper in half. Turn the folded edge closest to you. With scissors cut 1 line from the folded edge to the middle. On one side of the cut, fold that layer to make a triangle in a V-fold. Do the same for the other side of the cut. This will be the mouth or beak of the creature.

2. Unfold the triangles and then unfold the entire piece of paper. Use your thumb to push the V-fold back as you fold the card shut again. Now the folds should go on the inside of the card and leave a space shaped like a big triangle.

3. Use a pencil and draw two short equal lines on the fold on the inside of the big triangle. Hold the paper by the main fold and with scissors cut along the two lines, through both sides of the card. Be careful not to cut too far. Keep them short so you don't cut the inside

| Worksheet 2 | Name: _____ |

fold. Fold one of these small layers back. Flip the card over and fold the small layer back.

4. Unfold the page and push the new small layers forward.

5. Fold another piece of paper in half. Open up this new card and line up the centre and the edges with your first piece of paper. Glue this piece to the back of your first piece of paper keeping the edges lined up.

6. On the other side, draw and colour your creature, adding details like eyes, a tongue inside the mouth and special features.

What I Observe

The name of my creature is _____ .

The big cut in the middle of the page became the _____ of my creature.

I used the following colours for my creature:

_____

_____

_____

Worksheet 3   Name: _____

4. fold

3. cut

5. fold

2. fold

1. cut

| Worksheet 4 | Name: _____ |

Building a Toy Animal

You'll need:

- paper towel tube cut to size, or a toilet paper tube
- coloured paper – green, red, yellow, brown, black, white
- glue or tape
- a pencil crayon, crayon or marker

What to do:

1. Take the rectangle of green paper. Put glue on one side. Then line up the paper with your paper tube and roll the glue side all around the paper tube.

2. Cut the yellow piece of paper. Put glue on one side and attach the yellow paper on top of the green paper.

3. Cut out the arms and the feet. Fold a tab on each arm on the shoulder edge. Put glue on the tab and attach the arms on the left and right sides of the yellow piece of paper. Fold a tab on the back heel edge of each foot. Put glue on the tab and attach the feet to the bottom of your figure so that the feet come out the front.

4. Cut out circles from the white paper for the first part of the eyes. Put glue on one side of the circles. Attach the eyes on the top part of your figure. Cut out circles from the black paper for next part of the eyes. Put glue on one side of the circles and attach the black circles on top of the white circles.

| Worksheet 5 | Name: _____ |

5. With a black marker or pencil or crayon, draw below the eyes 2 dots for the nose and draw a V-shaped line for a mouth.

6. Cut a thin strip of red paper for the tongue. Curl one end of the paper by rolling the paper up. Put glue on the other end of the red paper. Attach the tongue to just below the mouth of your figure.

What I Observe

The name of my frog is _____ .

What are some things you can do with your toy frog?

_____

_____

_____

> Teacher Notes

BUILDING A WATER WHEEL AND A BOAT

LEARNING INTENTION:

Students will use problem solving skills, knowledge and comprehension in following instructions to design and build a water wheel. Students will learn how to apply knowledge from previous investigations and choose appropriate materials and designs in order to build a model boat.

SUCCESS CRITERIA:

- design and develop plans, considering properties of materials in decisions and plans
- make and record observations of the final product by comparing and sharing similarities, differences and benefits of designs.
- describe the purpose behind each part of the objects built.

MATERIALS NEEDED:

- *Building a Water Wheel* Worksheet 1 and 2 for each student
- *Building a Boat* Worksheet 3 and 4 for each student
- ruler, pencils or markers
- plastic or styrofoam dinner plates, small and large
- plastic cups, 1-2 ounces
- plastic cups, regular size drinking cups
- round wooden skewers or pieces of dowel
- stapler, wide masking tape, glue, string or rubber bands
- object that will make a hole (scissors, compass point, metal skewer, or even a ball point pen would work but then you have to make the hole big enough for the round wooden dowel or skewer to turn easily)
- water, water trench or bucket, sink and tap, or a clear aquarium of water
- popsicle sticks
- aluminum foil
- *optional* – foam pool noodle toy (cut into 3 cm thick cylinders), or corks, a 500ml plastic water bottle cut lengthwise in half (like the hull of a boat)
- plastic straws, cut in half for sizing

PROCEDURE:

Ideas from following websites, with some photos for visuals:

- http://goo.gl/DTL0sT
- http://goo.gl/9r4bGi
- https://goo.gl/iNBPdD

1. Start a discussion with students about wheels. What do wheels look like? How do they move? Ask students if they ever heard of a water wheel. Explain that water makes a water wheel turn round and round.

2. With the class, read through the Introductory Question, Materials need and What to Do sections on the Building a Water Wheel Worksheet. Demonstrate each step so that students will be prepared. (Build a water wheel beforehand, or the parts so that students can refer to each step. Read through the What I Observe section of the Worksheet to prime students with what to look for. Also demonstrate how to hold the water wheel during the experiment and how to pour water in line with the cups so that students can successfully recreate results.

3. Students will continue by making their own water wheel independently or in pairs. Set up a station with a sink and tap or with a bucket and water for the experiment.

4. Ask students if they have ever been on a boat. Did the boat float on top of the water? What was the boat made of?

5. Demonstrate how a plastic tray (or styrofoam tray) floats on the surface of water.

6. Tell students they have a chance to design and build a boat. But the boat must float. Each student will start with a plastic or styrofoam tray.

7. With the class, read through the Introductory Question, Materials, What to Do and My Design sections on the Building a Boat Worksheet. Read the What I Observe section of the Worksheet so that students know what to look for. When students have completed their design they can build their boat. (If students used certain kinds of glue, you may want to do Step 8 the next day.)

Teacher Notes

8. Set up a station with a sink and tap or with a bucket and water so that students can test their boats. Allow students to change their design and fix their boats as needed. Give students time to record the observations on the Worksheet or communicate their findings in a class discussion.

DIFFERENTIATION:

Slower learners may benefit from having the 3 lines already drawn on the bottom side of their water wheel plates (Step 1 of What I Do, Building a Water Wheel Worksheet). This project can also be done as part of a garden unit. Step 5 of What to Do can be done in a grassy area or garden spot as a way to water plants.

For enrichment, learners could investigate making water wheels out of other items, such as spoons, popsicle sticks or other items. Also, ask learners to investigate what might happen if the cups in the water wheel are fastened on an angle rather on the edge of the plate. Would more water or less water go in the cup? Would this make a difference to the speed or use of the water wheel? As well, learners can investigate how the motion of the water wheel could be applied to something else, such as a wheel or gear or fan, to make it move.

| Worksheet 1 | Name: _____ |

Building a Water Wheel

How can water can make things move? Let's build a Water Wheel to investigate how water makes things move. You will need a partner for this project.

You'll need:

- 2 plastic or Styrofoam dinner plates
- 6 plastic cups
- round wooden skewers or pieces of dowel
- stapler or wide masking tape
- object that will make a hole (scissors, compass point or a metal skewer
- water
- water trench or jug, sink and tap

What to do:

1. Use a ruler and draw a line on the bottom side of a plate from one edge of the plate through the centre to the other edge of the plate. Draw 2 more lines like this to make 6 segments around the edge of the plate. Find the centre point of the plastic plate. Carefully make a small hole through the centre point. Do this for both plates.

2. Staple or tape the plastic cups to one plate so that the cups line up with the lines drawn. Keep the open part of the cup along the edge of the plate.

| Worksheet 2 | Name: _____ |

3. Place the other plate on top of the cups and slide the wood skewer or dowel through the hole in each plate.

4. Staple or tape the second plate to the other side of the plastic cups. Look at every part of the water wheel to make sure it is ready for testing.

5. One partner will hold the water wheel by the skewer or dowel over a bucket or a sink. The other partner will carefully pour water into one of the cups of the water wheel.

What I Observe

What happened when I poured water onto the water wheel? _____

Did the water stay in the cups of the water wheel?

_____

_____

What does a waterwheel do?

_____

_____

_____

_____

Worksheet 3   Name: _____

Building a Boat

If you make a boat, you need to make sure it will float! What will you use to make your boat? Will you use plastic, paper, wood, foam, aluminum foil or something else?

You'll need:

- plastic or Styrofoam trays, small
- small sheets of paper
- popsicle sticks
- plastic straws
- rubber bands
- glue, masking tape or stapler
- foam or styrofoam pieces
- a bucket of water or a sink and tap

What to do:

1. Think about your plastic or Styrofoam tray. What will you add to make it look more like a boat? Draw a picture of your boat in the box on Worksheet 2.

2. Label each part of your boat. What material will you use?

3. Build your boat.

4. Test it to see if it will float. Make changes to your boat to make it better.

5. Complete the **What I Observe** part of the Worksheet.

Worksheet 4   Name: _____

My Design

What is the name that I gave my boat?
_____

What did I use to make my boat?
_____
_____

What I Observe

Did my boat float?   **Yes**   **No**
Did my boat stay together?   **Yes**   **No**

Find a partner and compare the boats you made.

What material did you use that was the same?
_____
_____

What did you use that was different?
_____
_____

Teacher Notes

TOPIC D: SENSES: EXPLORE YOUR SENSES

LEARNING INTENTION:
Students will learn about the five senses and how they are used to identify objects in our world.

SUCCESS CRITERIA:
- identify the five senses and explain their function
- give examples of how each sense is used to recognize objects in our environment
- gather and record data in diagrams and in charts
- make connections to your environment

MATERIALS NEEDED:
- a copy of *The Five Senses* Worksheet 1 for each student
- a copy of *Sensing the World Around Us* Worksheet 2 and 3 for each student
- a copy of *What's the Sense?* Worksheet 4 for each student
- a copy of *How Can You Sense It?* Worksheet 5 and 6 for each student
- a copy of *How Does It Feel?* Worksheet 7 for each student
- a copy of *How Does It Taste?* Worksheet 8 for each student
- an apple, chart paper, markers, pencils
- chocolate pieces, pretzels, lemon wedges, rhubarb pieces (a piece for each student)

PROCEDURE:
*This lesson can be done as one long lesson, or be divided into four shorter lessons.

1. Introduce to students our five different senses, ensuring their understanding/ meaning of each one. Discuss how we use our senses to identify objects/ the world around us. At this point, it would be beneficial for students to engage in a "knee to knee, eye to eye" activity where they turn and talk with a partner to brainstorm things that they can see, things that they can hear, things that they can smell, things that they can taste, and things that they can feel. Come back as a large group after each one to share responses orally. This will help to get ideas flowing. An option is to record responses on chart paper.

2. Give students Worksheet 1, 2, 3 and 4 to complete.

3. Explain to students that we sometimes use more than one sense to recognize things. Giving them an example of an object such as an apple (pass it around), ask them what senses they could use to identify it. Give students Worksheets 5 and 6 to complete.

4. Give students Worksheet 7 to complete.

5. Discuss with students how the tongue has different areas on it that are sensitive to different tastes (sweet, salty, bitter, and sour). An option is to give students a piece of a food to taste that will be detected by each area of their tongues. Have students determine which part of their tongue is sensing each food item. For example, give them a piece of chocolate, a pretzel, a lemon wedge, and a piece of rhubarb.

6. Give students Worksheet 8 to complete.

DIFFERENTIATION:
Slower learners may benefit by listing only two or three things per sense on Worksheets 2 and 3. An additional accommodation is to work in a small group with teacher support to read clues on Worksheet 4.

For enrichment, faster learners could come together as a group to share their own ideas from Worksheet 5. Using markers, they can display their ideas that they came up with on chart paper and then present it to the larger group. A further enrichment activity would be to have these students create clues of their own and have a partner match it to one of the senses.

Worksheet 1   Name: _____

The Five Senses

Use the words in the word box to complete the sentences.

```
────────────── Word Box ──────────────
   tongues    ears    noses    see    smell
   hear    skin    eyes    taste    feel
```

1. We _____ with our _____.

2. We _____ with our _____.

3. We _____ with our _____.

4. We _____ with our _____.

5. We _____ with our _____.

Worksheet 2   Name: _____

Sensing the World Around Us

Our senses allow us to learn, to protect ourselves, and to enjoy our world. The five senses are taste, touch, smell, sight, and hearing.

Make a list of things that you can sense with your tongue, touch, nose, eyes, and ears.

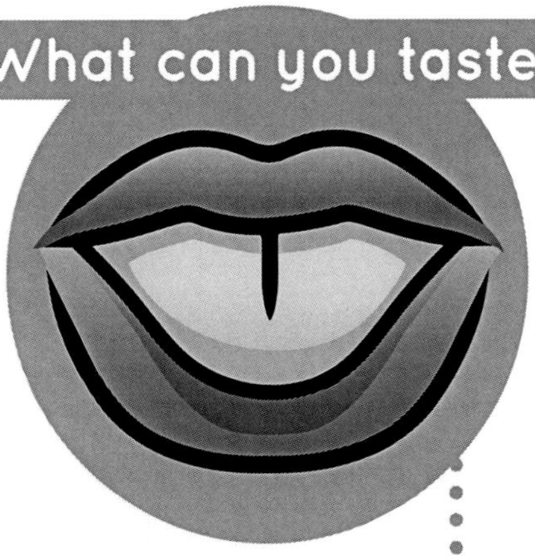

What can you taste?

_____

_____

_____

_____

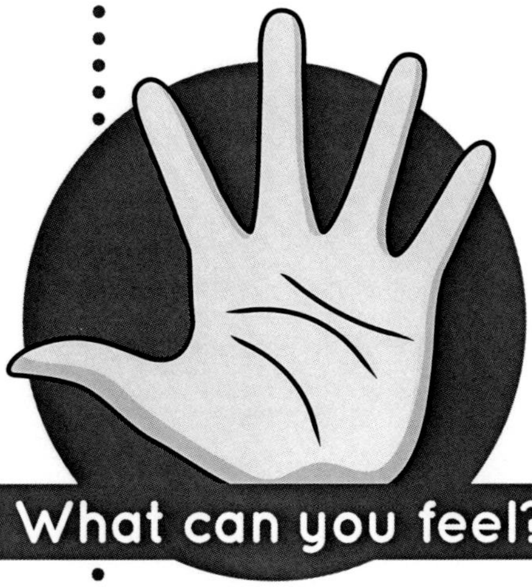

What can you feel?

_____

_____

_____

_____

Worksheet 3   Name: _____

What can you smell?

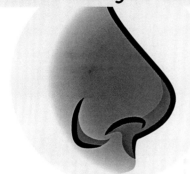

_____

_____

_____

_____

_____

_____

_____

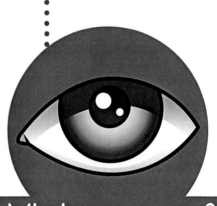

What can you see?

_____

_____

_____

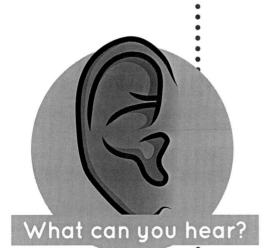

What can you hear?

| Worksheet 4 | Name: _____ |

What's the Sense?

Fill in the blanks to the clues using the sense words from the Word Box.

Word Box

see   hear   feel   smell   taste

Clues:

I know that cotton candy is sweet because I can _____ it.

I know that birds chirp because I can _____ them.

I know that a rock is hard because I can _____ it.

I know that flowers are fragrant because I can _____ them.

I know that ice is cold because I can _____ it.

I know that thunder is loud because I can _____ it.

I know that the sun is bright because I can _____ it.

I know that a lemon is sour because I can _____ it.

Worksheet 5   Name: _____

How Can You Sense It?

We can use more than one of our senses at a time to recognize things.

In the chart, list the senses that you **could** use to recognize what each object is. Then, add two of your own ideas.

Object	Senses Used
Popcorn	
Guitar	
Jello	
Rain	
_____	
_____	

Worksheet 6

Name: _____

You can use more than one sense to recognize these objects. But, can you sort them by using these clues?

1. Draw a **blue** circle around three things that you can **hear**.

2. Draw a **red** circle around three things that have a strong **smell**.

3. Draw a **green** circle around two things that are a treat to **eat**.

4. Draw a **purple** circle around one object that you **cannot** hear, smell, or taste.

Worksheet 7   Name: _____

How Does It Feel?

We can use any part of our skin to touch. By using our sense of touch, we can feel if something is hard or soft, wet or dry, or hot or cold.

Sort these objects by how they feel to the touch.

sponge	velvet	radiator	sandpaper
snow	a table	fur	hot chocolate
wood	fire	a rock	a cotton ball
ice cream	a brick		lemonade

Hard	Soft	Cold	Hot

Challenge! Add some words of your own to fill up the table.

Worksheet 8   Name: _____

How Does It Taste?

Did you know that different parts of your tongue are sensitive to different tastes? There are five taste zones on the tongue. They are sweet, bitter, salty, sour, and savory.

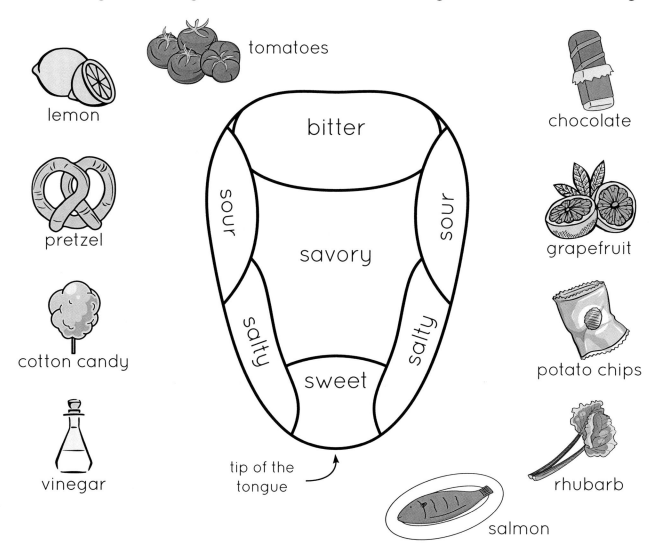

Challenge!

1. Colour the **sweet** area orange. Colour the **bitter** area **blue**. Colour the **savory** area **red**.

2. Colour the **salty** areas green and the **sour** areas **purple**.

3. Draw a line from the food words to the area on the tongue that you would taste them.

Teacher Notes

APPLYING YOUR SENSES

LEARNING INTENTION:
Students will learn how the senses are applied to describe and sort things in our environment.

SUCCESS CRITERIA:
- use the senses of sight, hearing, and smell to identify objects in a local area
- use the senses of touch, smell, taste, and hearing to recognize familiar objects
- memorize a group of objects that are within sight
- choose and use a sense to sort a group of objects
- use pictures and words to record responses
- make connections to the environment

MATERIALS NEEDED:
- a copy of *Getting a Sense of the Area* Worksheet 1 and 2 for each student
- a copy of *Grab Bag of Sensible Fun* Worksheet 3, 4, 5, and 6 for each student
- a copy of *How Sharp is Your Memory?* Worksheet 7 for each student
- a copy of *Sort by Senses* Worksheet 8 for each student
- 10 – 12 pairs of children's sunglasses (with masking tape covering the lenses)
- 6 timers, 6 large sheets of newspaper, chart paper, markers, clipboards, pencils

For touching activity:
- 6 grab bags of assorted objects (recognizable to the touch) such as a rock, paper clip, golf ball, a water bottle cap, eraser, a pine cone, a battery, etc.

For smelling activity:
- 4 – 6 plastic containers with tiny holes in lids, each containing coffee grinds
- 4 – 6 plastic containers with tiny holes in lids, each containing lemon wedges
- 4 – 6 plastic containers with tiny holes in lids, each containing onion slices
- 4 – 6 plastic containers with tiny holes in lids, each containing cinnamon

For tasting activity:
- 4 – 6 plastic containers with lids, each containing a few apple slices
- 4 – 6 plastic containers with lids, each containing a few onion slices
- 4 – 6 plastic containers with lids, each containing some shredded carrot
- 4 – 6 plastic containers with lids, each containing melon pieces

For listening activity:
- 4 – 6 plastic containers with lids, each containing coins
- 4 – 6 plastic containers with lids, each containing uncooked rice
- 4 – 6 plastic containers with lids, each containing small rocks
- 4 – 6 plastic containers with lids, each containing uncooked macaroni

For visual activity:
- 6 grab bags of 15 assorted small objects such as a rock, paper clip, golf ball, a water bottle cap, eraser, a pine cone, a battery, a pencil, a marble, a lollipop, a whistle, a spoon, popsicle stick, a juice box, an aromatic marker

PROCEDURE:
*This lesson can be done as one long lesson, or done in seven or eight shorter lessons.

1. Give students Worksheet 1, and a clipboard and pencil. Take them on a walk of the neighbourhood. Instruct them to use their senses of sight, hearing, and smell to observe things in the area. An option upon returning to the classroom is to divide students into small groups to orally share their observations. Next, give students Worksheet 2. Take them on a walk in a local park or forest area. Instruct them to use their senses of sight, hearing, and smell to observe things in this area. Upon returning to the classroom, have students orally share their observations. Pose these questions to the large group:

Teacher Notes

- *Were any of the sights, sounds, smells the same in both the neighbourhood and park/forest?*
- *Were there any differences in the sights, sounds, smells in these areas?*

2. Divide students into small groups of 4. Give each group a grab bag of assorted small objects and a pair of blinded sunglasses. Students will take turns to pull two objects out of the bag. Instruct them to use their sense of touch to describe it first, and then name it. They will draw and label their objects on Worksheet 3.

3. Students will continue to work in groups of 4. Give each group 4 plastic containers (each with different contents) that have small holes punched through the lid to allow aroma out (use empty margarine containers and punch holes in the lids with a small nail and hammer). Students will take turns to smell the objects in the containers. Instruct them to use their sense of smell to describe it first, and then name the object inside. They will draw and label their objects on Worksheet 4.

   *Materials suggested can be substituted for other familiar smelling things.

4. Students will continue to work in groups of 4. Give each group 2 pairs of sunglasses and 4 plastic containers (with lids, no holes), that have different food objects in them.

   Students will take turns (in pairs) to taste the objects in 2 of the containers. Instruct them to use their sense of taste to describe it first, and name the object inside. They will draw and label their objects on Worksheet 5. *Materials suggested can be substituted for other safe, familiar tasting foods. It is important to check before beginning this activity for any food allergies students may have. Also, it is important to explain to students that while you are encouraging them to use their sense of taste for this activity, it is not safe to do so outside of this activity. It is important to always check first with a responsible adult.

5. Students will continue to work in groups of 4. Give each group 4 plastic containers (each with different non-food related contents). Students will take turns to listen to the sound the objects in the containers make as they are shaken or moved around. Instruct them to use their sense of hearing to describe it first, and then name the object inside. They will draw and label their objects on Worksheet 6. *Materials suggested can be substituted for other familiar sounding things.

6. Students will continue to work in small groups. Give each group a bag of 15 assorted small items, a timer, and a clipboard and pencil. They will empty the contents of the bag and give themselves 2 minutes to memorize what they see. Once the time is up, the objects are to be covered while students record from memory the objects that they saw, using Worksheet 7.

   *Materials suggested can be substituted for other familiar objects.

7. Give students Worksheet 8. Using the objects from the grab bag in the previous activity, students will sort them according to one or more of the senses. A follow-up option is to have students share their sorting rule with a classmate to compare different ways to sort.

Differentiation:

Slower learners may benefit by working as a small group and having a grab bag of only 10 assorted objects. An additional accommodation would be to work as a small group with teacher support to complete Worksheet 8.

For enrichment, faster learners could make a list of visual clues that describe a classmate. These lists could be read to the large group, who will make a guess of who the clues are about.

Worksheet 1

Name: _____

Getting a Sense of the Area

Take a walk around your neighbourhood. Get a sense of the area by telling about what you see, hear, and smell.

In the _____ :

I see:
_____
_____
_____
_____

I smell:
_____
_____
_____
_____

I hear:
_____
_____
_____
_____

Worksheet 2

Name: _____

Getting a Sense of the Area

Now take a walk around a local park or forest. Get a sense of this area by telling about what you see, hear, and smell.

In the _____ :

I see:
_____
_____
_____
_____
_____

I smell:
_____
_____
_____
_____
_____

I hear:
_____
_____
_____
_____
_____

| Worksheet 3 | Name: _____ |

Grab Bag of Sensible Fun!

Activity #1: Touch and Guess

1. Put the blinded sunglasses on.

2. Reach into the grab bag and pull out one object.

3. Use your sense of touch to **describe** the object.

4. Guess what the object is.

5. Take off the sunglasses and check your guess. Draw and label the object in the box below.

6. Repeat steps 2, 3, 4, and 5.

7. Pass the grab bag and sunglasses to another group member.

Object #1	Object #2

| Worksheet 4 | Name: _____ |

Activity #2: Smell and Guess

1. Use your sense of smell to **describe** the object inside container #1.

2. Guess what the object is. Pass the container to another group member.

3. After each group member has described and made a guess about what the object is, open the lid of the container to check your guess.

4. What is the object? Draw it and label it in the box below.

5. Repeat all the steps for each container on the table.

Object #1	Object #2

Object #3	Object #4

Worksheet 5

Name: _____

Activity #3: Taste and Guess

1. Two group members will put blinded sunglasses on. Another group member will open the lid of a tasting container.

2. The two group members will plug their noses and reach into the container to take out an object of food.

3. Use your sense of taste to **describe** the food.

4. Guess what the food is.

5. Take off the sunglasses and check your guess. Draw and label the food object in the box below.

6. Repeat all the steps using another tasting container.

7. Switch roles with the other two group members so they can become taste testers!

Object #1	Object #2

_____   _____

| Worksheet 6 | Name: _____ |

Activity #4: Listen and Guess

1. Use your sense of hearing to **describe** the object inside container #1.

2. Guess what the object is. Pass the container to another group member.

3. After each group member has described and made a guess about what the object is, open the lid of the container to check your guess.

4. What is the object? Draw it and label it in the box below.

5. Repeat all the steps for each container on the table.

Object #1

Object #2

Object #3

Object #4

Worksheet 7   Name: _____

How Sharp is Your Memory?

It is time to play a memory game!

You'll need:

- a grab bag of 15 objects
- a table
- a pencil
- a large sheet of newspaper -a timer

What to do:

1. Empty out the grab bag of objects onto the table.
2. Set the timer for 2 minutes. Try to remember the objects that you see.
3. When the time is up, cover the objects with the sheet of newspaper.
4. In the box below, record the objects that you saw.

Worksheet 8　　　Name: _____

Sort by Senses

Use the grab bag of objects from Worksheet 7.

Show how you can sort them.

What sense will you use to sort?

To sort the objects, I used the sense of _____.

Teacher Notes

ANIMAL SENSES

LEARNING INTENTION:
Students will learn about how animals use their senses to survive in the natural world.

SUCCESS CRITERIA:
- recognize that animals need to use their senses to survive in nature
- research and describe how an animal uses its senses
- orally communicate research findings with other classmates
- listen to and record interesting facts about how animals use their senses to survive

MATERIALS NEEDED:
- a copy of *Sensing to Survive!* Worksheet 1 and 2 for each student
- a copy of *Making Use of Its Senses* Worksheet 3 and 4 for each student
- a copy of *Sensing the Animal Facts!* Worksheet 5 for each student
- access to the internet, or local library
- chart paper, markers, pencils, clipboards

PROCEDURE:
*This lesson can be done as one long lesson, or done in four or five shorter lessons.

1. Using Worksheets 1 and 2, do a shared reading activity with the students. This will allow for reading practice and learning how to break down word parts in order to read the larger words in the text. Along with the content, discussion of certain vocabulary words would be of benefit for students to fully understand the passage.

   Some possible vocabulary words to focus on are: *prey, texture, tentacles, sensors, scan, predators, echo, chemoreceptors, nectar, perched, tongue, echolocation, chemicals*

2. Assign students to one of the five different animals mentioned in Worksheet 1 and 2 (snake, bat, star-nosed mole, butterfly and eagle). Give students Worksheets 3 and 4, and a clipboard and pencil. They can visit the local library or access the internet to gather further information to help them find out how their chosen animal uses its senses to help it survive in nature. *An option is to choose different animals for students based on previous work they have done.

3. Divide students into groups of five, each group consisting of a student representing one of the 5 animals mentioned in Worksheet 1 and 2. Each student in the group will orally present their research on how their chosen and unique animal uses its senses to survive in nature. Instruct students to record one interesting fact as each group member presents their information. Come back as a large group to share some of the facts students heard about, in order to promote some rich discussion and the usage of scientific content words.

4. As an activity to enhance the learning about animal senses, show students Wild Kratts episode Platypus Café. Episode can be accessed at www.youtube.com

   *In this episode, the platypus uses electrosense to detect the location of objects. The rattle snake uses heat sense to see objects more clearly.

DIFFERENTIATION:
Slower learners may benefit by working as a small group with teacher support and direction to complete Worksheets 3 and 4, using one animal. Sections of the assignment could be divided between these students to lessen the work load.

For enrichment, faster learners could illustrate how the platypus and the rattle snake each uses their "extra sense" to survive in nature. Encourage students to add a descriptive sentence.

Sensing to Survive!

Did you know that animals have senses too? They use their senses to help them survive in nature. Let's learn more about this!

Fast Facts!

Did you know that a snake uses its tongue to "taste" the air? When a snake flicks out its tongue, it is able to sense smells in the air. When the snake's tongue goes back into its mouth, it touches the top of its mouth and tells the snake what it smells. A snake uses its tongue to smell prey and predators that are close by.

Have you ever heard of **echolocation**? Echolocation is what bats use to find their way in the dark and to find food. Bats send out sound waves with their mouth or nose. When a sound wave hits an object, an echo comes back. This helps the bat to know what the object is, where it is, and how close it is.

Bats can tell the size, shape, and texture of an insect all from its echo.

Worksheet 2   Name: _____

The star nose mole is an animal that has a great sense of touch. This is because it has 22 tentacles that form a star shape on its nose. The star nose mole cannot see well. It uses its tentacles to feel around and to find its prey like worms, small fish, and insects. A star nose mole can touch as many as 12 different objects per second!

Did you know that a butterfly tastes with its feet? A butterfly's feet have sensors that can taste the nectar in a plant. If the nectar tastes good, the butterfly will eat it. A female butterfly tastes a plant to find out if it is a good place to lay eggs. She uses the chemoreceptors on her legs to find the right match of plant chemicals. When she finds the right plant, she lays her eggs.

Birds of prey such as eagles and hawks have very good eyesight. They can scan the Earth as they are flying or perched, and spot their prey from far away.

| Worksheet 3 | Name: _____ |

Making Use of Its Senses

Take a look back at Worksheet 1 and Worksheet 2. Use your worksheets, visit a library or use a computer to find out how your animal uses all 5 of its senses to survive in nature.

The _____

uses its **sense of sight** to_____

_____

_____

_____

_____

The _____

uses its **sense of hearing** to _____

_____

_____

_____

_____

Worksheet 4   Name: _____

The _____
uses its **sense of smell** to _____

_____

_____

_____

_____

The _____
uses its **sense of taste** to_____

_____

_____

_____

_____

The _____
uses its **sense of touch** to _____

_____

_____

_____

_____

| Worksheet 5 | Name: _____ |

Sensing the Animal Facts!

Getting the Facts!

Listen to your group members as they tell about how their animal uses its senses to survive in nature. Record some facts that you learned about.

1. _____
   _____
   _____
   _____

2. _____
   _____
   _____
   _____

3. _____
   _____
   _____
   _____

4. _____
   _____
   _____
   _____

Teacher Notes

PROTECTING THE SENSES

LEARNING INTENTION:
Students will learn about the protective parts of the eye and ear; and about the importance of protecting our senses that contribute to our safety in daily living activities.

SUCCESS CRITERIA:
- identify and label some parts of the eye, and determine their use in protecting the eye
- identify some parts of the ear and determine their use in protecting the ear
- describe the hearing ability of an animal
- recognize some parts of the nose and their purpose
- recognize ways that we can protect our senses of taste, smell, and touch
- describe the affects of losing a sense has on daily living
- recognize ways that the five senses contribute to our daily lives

MATERIALS NEEDED:
- a copy of *The Eye* Worksheet 1 and 2 for each student
- a copy of *The Ear* Worksheet 3 and 4 for each student
- a copy of *Honing in on Hearing* Worksheet 5 and 6 for each student
- a copy of *Being Nosey About the Nose* Worksheet 7 and 8 for each student
- a copy of *Protecting the Others* Worksheet 9 for each student
- a copy of *Sensible Safety!* Worksheet 10 for each student
- about 10 – 12 small hand mirrors
- access to music
- access to the internet, or local library
- access to television/ video machine and video
- 10-12 pairs of children's sunglasses (with masking tape covering the lenses)
- different grab bags of small assorted items (one for each student)
- chart paper, markers, pencils, pencil crayons, clipboards

PROCEDURE:
*This lesson can be done as one long lesson, or done in eight or nine shorter lessons.

1. Give students Worksheet 1 and a small hand mirror. They will take a close look at their own eye. They will draw a diagram of their eye, labelling its basic parts using words from the Word Box.

2. Using Worksheet 2, do a shared reading activity with the students. This will allow for reading practice and learning how to break down word parts in order to read the larger words in the text. Along with the content, discussion of certain vocabulary words would be of benefit for students to fully understand the passage.

   Some interesting vocabulary words to focus on are: **eyebrows, eyelids, eyelashes, pupil, iris, protect, moist, sweeping, centre, muscle, sweat, blinking, harm**

3. Discuss with students the importance of protecting their eyes and ways that they can actively do that (e.g. wear sunglasses, wear corrective lenses if needed, read with the right amount of light, get regular eye exams, wear protective sporting eye gear).

4. Give students Worksheet 3. As they listen to music, they will experiment by covering/ cupping the ear in different ways. Students can discuss the effects of the size and shape of the ear. (The shape of the outer ear helps to funnel sound to the inner parts of the ear, then nerve impulses send messages to the brain to interpret the sound.

Teacher Notes

5. Using Worksheet 4, do a shared reading activity with the students. This will allow for reading practice and learning how to break down word parts of larger words in the text. Along with the content, discussion of certain vocabulary would benefit students.

   Some interesting vocabulary words to focus on are: *outer ear, protected, separates, sound, vibrations, ear canal, middle ear, inner ear, nerve impulses, collects, ear drum, cochlea, curled*

6. Discuss with students the importance of protecting their ears and ways that they can actively do that (e.g. wear ear plugs when in areas with excessive noise, listen to music or television at an appropriate volume, clean ears regularly, get medical check-ups).

7. Give students Worksheet 5. They will circle the animals that have a good sense of hearing, and explain their choices. (All animals should be circled except the hippopotamus and the turtle.)

8. Give students Worksheet 6. They can visit the local library or access the internet to gather information on the hearing ability of their chosen animal.

9. Using Worksheets 7 and 8, do a shared reading activity with the students. This will allow for reading practice and learning how to break down word parts of larger words in the text. Along with the content, discussion of certain vocabulary would benefit students.

   Some interesting vocabulary words to focus on are: *nostrils, sinuses, healthy, organ, blood vessels, septum, throat, moistens, mucous, mucous membrane, trachea, breathe, lungs, nasal cavity*

10. Divide students into pairs and give them Worksheet 9. They will engage in a 'think, pair, share' activity to discuss and record things we can do to protect our senses of touch, taste, and smell. *Inform students that we touch with all parts of our body (via our skin). An option is to come back as a large group to share their responses.

11. To teach students what it would be like to not be able to see or hear, but still have to function using other senses, try these activities:

    - have students put the blinded sunglasses on and sort a bag of items (having to determine their own sorting rule)
    - have students watch a television show/video without the volume on, afterwards have them try to discuss the events, then watch it with the volume on to see how accurate they were

12. Engage students in a discussion about ways that people adapt to the loss of a particular sense. Also discuss if there are any limitations to our senses. (They should be able to draw upon their own experiences from the previous activities.)

13. Give students Worksheet 10 to complete.

DIFFERENTIATION:

Slower learners may benefit by working with a strong peer in order to 'navigate' the research required to complete Worksheet 6.

For enrichment, faster learners could create a poster that details the importance of protecting our senses.

Worksheet 1

Name: _____

The Eye

Use a small mirror to take close look at your eye.

Take a Look!

1. What parts do you see?

2. Draw your eye.

3. Use the words in the Word Box below to label some of its parts.

---- Word Box ----

eyebrow   **eyelid**   **eyelashes**   **pupil**   **iris**

Worksheet 2   Name: _____

Have you ever wondered what these parts of your eye do? They protect the eye. Let's learn more about this!

Eyebrows protect our eyes from water or sweat that could drip down and come into our eyes.

Eyelashes protect our eyes by sweeping away dust and sand that can get into the eyes and harm them. Eyelashes also help to keep sweat or rain out of the eyes.

Eyelids protect our eyes by opening and closing over them to keep them moist. This is called blinking.

The pupil is the centre of the eye.

When there is a lot of light, the pupil gets smaller to protect the eye from getting too much light in it.

When there is not a lot of light, the pupil gets bigger so that it can let in as much light as possible to give the eye more seeing power.

The iris is the colour part of the eye.

The iris is like a muscle for the eye. It helps the pupil to grow bigger, or get smaller.

It works with the pupil to protect the eye from getting too much light in it, and it works to let more light in if the eye needs more seeing power.

Worksheet 3 Name: _____

The Ear

Listening In

Take a close look at the ear. Why do you think it is shaped the way it is?

I think the ear is shaped this way because

_____

_____

_____

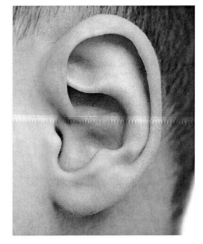

Try this!

Your teacher will play some music.

1. Cup your hand in front of your ear. Listen to the music.

2. Cup your hand behind your ear. Listen to the music.

3. With a classmate, talk about what happened.

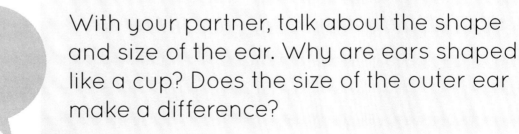

With your partner, talk about the shape and size of the ear. Why are ears shaped like a cup? Does the size of the outer ear make a difference?

Worksheet 4   Name: _____

Our ears have three main parts. They are the outer ear, the middle ear, and the inner ear. Each part does something to protect the ear. Let's learn more about this!

The **outer ear** is the part that you can see. It opens into the ear canal. The outer ear collects sounds and it is like a path for the sound to get to the inside of the ear.

The **inner ear** has a small curled tube in it called the cochlea.

Inner Ear

Outer Ear

The **cochlea** has tiny hairs that move. They turn the vibrations sent by the middle ear into nerve impulses that go to the brain. The brain tells you what you are hearing.

Middle Ear

The **ear drum** separates the outer ear from the middle ear.

The **middle ear** has three small bones in it that move to send sound vibrations to the inner ear.

The **ear canal** is protected by tiny hairs. It makes wax that traps dirt and dust from getting into the ear.

Worksheet 5 Name: _____

Honing in on Hearing

Look at the ears of the animals in the pictures. Circle the ones that you think have the best hearing.

Why do you think the animals you circled have a good sense of hearing?

_____

_____

_____

| Worksheet 6 | Name: _____ |

Choose one of the animals you circled on Worksheet 5. Visit a library or use a computer to find out more about this animal's sense of hearing.

The _____

I learned that _____

_____

_____

_____

I also learned that _____

_____

_____

_____

Draw a picture to go with one of the facts you learned about this animal.

Being Nosey About the Nose

Your nose is the organ that lets you smell, and it helps you taste because it smells the food you are eating. The nose has another job too. It breathes in air to your lungs that are inside your body. Let's learn more about the parts of your nose that help you to do these things!

The **nostrils** are the openings to your nose. They breathe in air that travels to your **nasal cavity**. From there, air goes down your throat into the **trachea**, which is like a windpipe going to your lungs. Then air fills your lungs so your body can breathe.

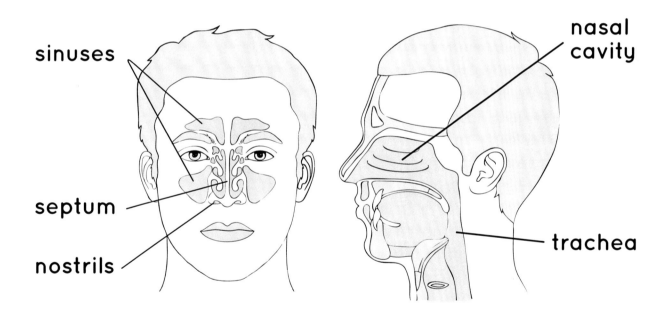

Did you know?

The inside of your nose has mucous membrane. When you breathe in, the membrane warms and moistens the air. The membrane and the tiny hairs inside your nose trap dust and dirt that would not be healthy for your lungs. Sinuses in your nose store mucous.

Question:

Why does my nose run?

Answer:

If you have a cold, your nose makes extra mucous to keep the germs from getting into your lungs. It could make so much that it runs out of your nose!

And did you know that if you are outside on a cold day, your nose warms up the cold air before it gets to your lungs? The blood vessels in your nostrils get bigger to warm the air, but this extra blood flow makes more mucous that runs out of your nose!

Question:

Why does my nose bleed?

Answer:

When you have a cold, you sneeze and blow your nose. Sometimes this could make the tiny blood vessel in your nose break and your nose might bleed.

And did you know that when the indoor air is very dry, the inside of your nose gets crusted and itchy. When you scratch that itchy nose, it may start to bleed.

An injury to the nose could also make it start to bleed. So take care and wear protective gear when playing sports!

Worksheet 9

Protecting the Others

Think, Pair, Share

With a partner, do some thinking and sharing of ideas about the ways we can protect our senses of touch, taste, and smell. Record your ideas.

Taste

Smell

Touch

Worksheet 10   Name: _____

Sensible Safety!

You have learned a lot about the five senses, and how important it is to protect them.

Now, illustrate ways that our five senses keep us safe and help us in our daily lives.

Teacher Notes

TOPIC E: NEEDS OF ANIMALS AND PLANTS: THE ANIMAL WORLD

LEARNING INTENTION:
Students will observe, describe, compare and classify living things based on visible characteristics. Students will identify ways in which living things are valued.

SUCCESS CRITERIA:
- recognize the physical characteristics of of a variety of animals
- compare and sort a variety of animals according to the characteristics
- research and describe the characteristics, habitat, food needs, special adaptations of an animal and the valuable role the animal plays in the ecosystem

MATERIALS NEEDED:
- a copy of *What Do You Know About Animals?* Worksheet 1, 2, 3 and 4 for each group of students.
- a copy of *Describing An Animal* Worksheet 5 and 6 for each student
- access to the internet and a local library
- Bristol board – ½ piece per student or similar
- chart paper, makers, pencils, pencil crayons, scissors, modelling clay

PROCEDURE:
*This lesson can be done as one long lesson or divided into shorter lessons.

1. Divide students into small groups of 3 or 4. Give each group a copy of Worksheets 1, 2, 3 and 4. Instruct them to talk about things that they know about the animals on each of the cards. After students have discussed each animal, ask each student to share one of two things that they know about one of two of the animals on the cards. An option is to record these student responses on chart paper for reference for future lessons.

2. Students will continue to work within their groups. Give students scissors. They will cut out the animals cards on the Worksheets 1, 2, 3 and 4. Give students categories for which to sort the animal cards. This will allow them to compare the physical characteristics of the animals, and to discuss why they have them. Some categories are:

    animals with wings / no wings, walkers, crawlers, hoppers, flyers, slitherers, animals with bills / no bills, feathers, fur, skin, scales, swimmers / non-swimmers, 2 legs, 4 legs, 6 legs, no legs

3. Students will research an animal that they would like to know more about. Give them Worksheets 5 and 6. Students can access the internet or visit the local library to gather information to help them answer the questions on Worksheets 5 and 6. Upon completion, students can display their work on Bristol board which could be posted in the classroom or around the school.

*An alternative follow-up activity is to have students participate in a carousel activity to read and learn about their classmates' chosen animals.

DIFFERENTIATION:
Slower learners may benefit by working in a small group with teacher support to complete the research about an animal. A section of the research project could be assigned to each student in the small group, so that all sections are filled in to complete one final project. Each of the sections could be done on large chart paper or on one piece of Bristol board and then displayed in the classroom.

For enrichment, faster learners could use modelling clay to make a sculpture of their animal.

Worksheet 1   Name: _____

Group Talk!

With you group members, talk about the things that you know about each animal on the cards below.

Worksheet 2　　　　　　　Name: _____

Group Talk!

With you group members, talk about the things that you know about each animal on the cards below.

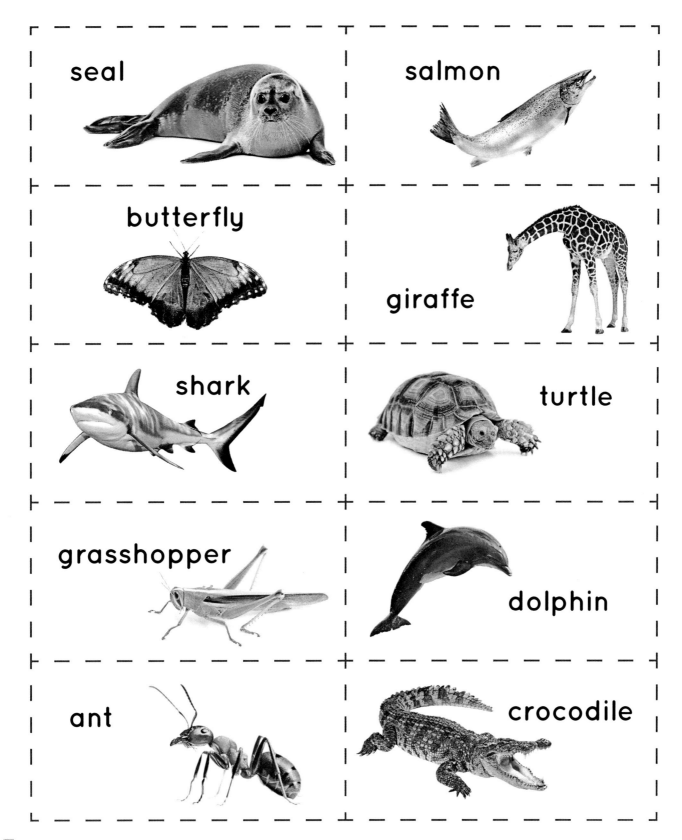

Worksheet 3   Name: _____

Group Talk!

With you group members, talk about the things that you know about each animal on the cards below.

Worksheet 4　　　　　　　Name: _____

Group Talk!

With you group members, talk about the things that you know about each animal on the cards below.

cow	pig
squirrel	lion
wolf	fox
raccoon	moose
duck	owl

Worksheet 5   Name: _____

Describing an Animal

Choose an animal that you would like to learn more about. Visit a library or use a computer to help you find out the answers to the questions below.

Let's Research!

Animal Name: _____

Where does this animal live?

_____

_____

_____

Make an illustration of this animal in its habitat.

| Worksheet 6 | Name: _____ |

What special adaptations does this animal have?

_____
_____
_____
_____

Why does this animal need these special adaptations?

_____
_____
_____
_____
_____

Draw some things that this animal eats.

With a partner, talk about why you think your animal is so interesting.

Teacher Notes

THE PLANT WORLD

LEARNING INTENTION:

Students will learn about the physical characteristics of plants and how they meet their needs.

SUCCESS CRITERIA:

- recognize a variety of plants in your neighbourhood
- identify the main physical characteristics of a plant
- compare a variety of plants according to their physical characteristics
- investigate the basic needs of a plant
- explain the function of the main physical characteristics of a plant
- describe how a plant uses its physical characteristics to meet its needs

MATERIALS NEEDED:

- a copy of *Plants in My Neighborhood* Worksheet 1 for each student
- a copy of *The Parts of a Plant* Worksheet 2 for each student
- a copy of *The Needs of a Plant* Worksheet 3, 4, and 5 for each student
- a copy of *The Sum of All Parts* Worksheet 6 and 7 for each student
- a copy of *Drinking Up the Nutrients!* Worksheet 8, 9, and 10 for each student
- 5 or 6 assorted plants
- 3 small potted plants for each group of students
- medium sized boxes (one for each group of students)
- access to water, a few measuring cups, a couple of watering cans
- a celery stalk with leaves, a tall glass of water, a magnifying glass, a spoon (a set for each pair of students)
- a few small bottles of red food colouring
- chart paper, markers, pencils, pencil crayons, clipboards, labels
- iPods or iPads (optional)

PROCEDURE:

*This lesson can be done as one long lesson, or be done in four or five shorter lessons.

1. Give students Worksheet 1, a clipboard and a pencil. Take them out into the neighbourhood to look for different kinds of plants. Encourage them to notice the types of trees, shrubs, grasses, flowered plants, etc. that are growing locally. An option is to give students iPods or iPads to take photos of the different vegetation that they see.

2. Display 5 or 6 assorted plants (roots exposed). Engage students in a discussion about the physical characteristics of each plant type. How they different? What do they all have in common? Ask students to look back at the plants that they drew on Worksheet 1 (or have taken photos of). How are they different? How are they the same? (Some common characteristics that should be noted are that all plants have a root system, stem, leaves. They differ in size, shape, colour, and some have flowers). Give students Worksheet 2. They will label the basic parts of a plant.

3. Give students Worksheets 3, 4, and 5. They will make a prediction about what may happen to a plant that is placed in a dark place, to one that is in sunlight but given no water, and to one that is in sunlight and given water. Over the next two weeks, students will make observations and conclusions about the growth of the plants and record them on Worksheet 4 and 5. (Sunlight, water, and air are needed to make plants grow.)

4. Divide students into pairs. Give them Worksheet 6. They will engage in a **think-pair-share** activity to discuss and then record answers to the questions on the Worksheet. A follow up option is to come back together as a large group to share responses. Student responses could be recorded on chart paper and displayed in the classroom for future reference.

5. Give students Worksheet 7. Read through with the students about the main parts of a plant and their purposes. Along with the content, discussion of certain vocabulary words would be of benefit to ensure students' understanding of the concepts.

Some interesting vocabulary words to focus on are: ***chlorophyll, attract, nutrients, absorb, pollinate, supports, anchor, minerals***

6. Divide students into pairs. Give them Worksheets 8, 9, 10, and the materials to conduct the experiment. Read through the materials needed and what to do sections to ensure their understanding of the task. Students will make observations before and after the experiment, then make conclusions and a connection to what they have already learned about the needs of plants.

*As an activity to enhance the learning about the physical characteristics and needs of plants, show students The Magic School Bus episode called "Gets Planted". Episodes can be accessed at www.youtube.com

DIFFERENTIATION:

Slower learners may benefit by working with a partner to make and record observations about the plants throughout the two weeks, or simply draw what they saw each week. An additional accommodation would be for these learners to work as one small group with teacher support to discuss the questions on Worksheet 6. Answers could be recorded on chart paper and posted in the classroom as reference throughout the unit.

For enrichment, faster learners could plant and care for the 5 or 6 assorted plants (used in item #2). They could chart their growth by measuring the plants' heights, or amount of flower production each has. This could be done by creating a simple bar graph.

Worksheet 1   Name: _____

Plants in My Neighbourhood

Have you ever noticed what kinds of plants are in your neighbourhood?

Take a walk around your neighbourhood. In the box below, draw and label the plants that you see.

Worksheet 2   Name: _____

The Parts of a Plant

Label the main parts of the plant below. Use the words in the Word Box to help you.

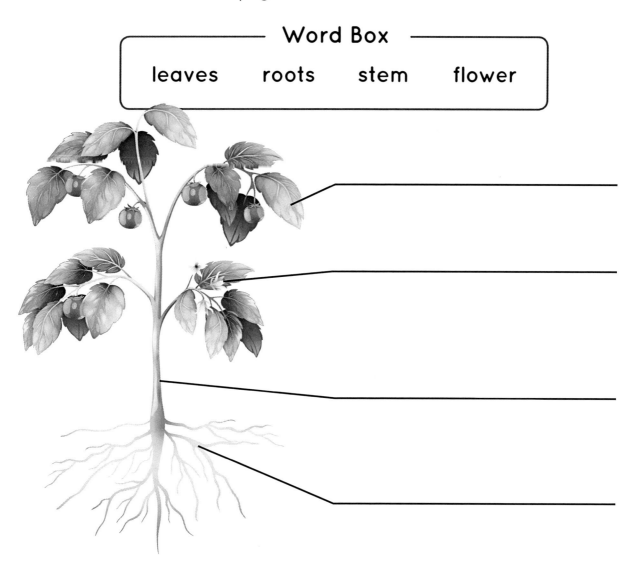

——— Word Box ———
leaves   roots   stem   flower

Challenge question:

Why are the roots of a plant important?

_____

_____

_____

| Worksheet 3 | Name: _____ |

The Needs of a Plant

Questions: What does a plant need to live and grow?

You'll need:

- 3 small potted plants (one labelled DARK, one labelled LIGHT, and one labelled LIGHT and WATER)
- water
- watering can
- measuring cup
- a medium sized box

What to Do:

1. Make a prediction about the answer to the question. Record it on your Worksheet.
2. Place the DARK plant under the box.
3. Place the LIGHT and the LIGHT and WATER plants in a sunny place.
4. Every 3 days, water the DARK plant, and water the LIGHT and WATER plant. Be sure to give them the same amount of water.
5. Record your observations of all plants at the end of each week, for 2 weeks.
6. Make conclusions about what you observed.

Worksheet 4   Name: _____

Let's Predict

What does a plant need to live and grow?

_____

_____

Let's Investigate

This is what I saw each week.

Week 1

	Dark Plant	Light Plant	Light and Water Plant
What colour are the leaves?			
Does the plant look healthy or sick?			

Week 2

	Dark Plant	Light Plant	Light and Water Plant
What colour are the leaves?			
Does the plant look healthy or sick?			

Worksheet 5   Name: _____

Drawing of the plant grown in the dark:

Drawing of the plant grown in sunlight, without water:

Drawing of the watered plant grown in sunlight:

Let's Conclude

What do plants need to live and grow?

_____

_____

How do you know this?

_____

_____

Worksheet 6   Name: _____

The Sum of All Parts

Think, Pair, Share

With a partner, do some thinking and sharing of ideas about the questions below. Record your ideas.

What does a plant use its roots for?

What does a plant use its stem for?

What does a plant use its leaves for?

What does a plant use its flowers for?

Worksheet 7 Name: _____

Leaves use energy from the sun to mix with their chlorophyll to make a sugar from the air. The sugar, air, and water that the plant takes in, makes food for it.

When **flowers** are pollinated, they produce fruit. Colourful flower petals attract animals and insects that help to pollinate the flowers. Flowers also make seeds that are needed to start new plants.

A **stem** of a plant supports it. The stem grows leaves that make food for the plant. The stem stores the food. The stem carries water and nutrients to other parts of the plant.

The **roots** anchor the plant to the ground. The roots of a plant absorb water and minerals that the plant needs to grow.

| Worksheet 8 | Name: _____ |

Drinking Up the Nutrients!

Do you want to see a plant have a drink of water to get the nutrients it needs to be a healthy plant?

You'll need:

- a celery stalk with its leaves on
- a tall glass of water
- a spoon
- a magnifying glass
- red food colouring

What to do:

1. Use the magnifying glass to take a look at the base of the celery stalk. On Worksheet 9, describe and draw what you see.

2. Next, put 5 drops of the red food colouring into the glass of water.

3. Stir the water and food colouring with the spoon.

4. Put the celery stalk into the glass of water. Leave it sit over night.

5. The next day, observe the celery stalk. Record your observations on Worksheet 9.

6. Make conclusions and a connection about what you have observed. Record them on Worksheet 10.

| Worksheet 9 | Name: _____ |

Let's Observe

Before beginning the experiment...

The base of the celery stalk looked _____
_____
_____
_____
_____

A drawing of the base of the celery stalk:

After doing the experiment...

The base of the celery stalk looked _____
_____
_____
_____

A drawing of the celery stalk:

The stem and leaves looked
_____
_____

| Worksheet 10 | Name: _____ |

Let's Conclude:

Why did the celery stalk and its leaves turn red?

_____
_____
_____
_____
_____
_____

With a partner, talk about how the structure and design of the celery stalk helps to get its nutrient needs.

Let's Connect It!

If you planted a garden, explain what your plants would need to grow and be healthy.

_____
_____
_____
_____
_____

Teacher Notes

NEEDS INTERTWINED!

LEARNING INTENTION:
Students will learn about what living things provide for each other, and the problems that could result from the loss of some kinds of living things.

SUCCESS CRITERIA:
- recognize that things used to meet needs are changed and returned in different forms
- identify what living things provided for other living things in order to have their needs met
- describe problems that could happen because of the loss of some kinds of living things

MATERIALS NEEDED:
- a copy of *Processing Needs* Worksheet 1 and 2 for each student
- a copy of *Give and Take* Worksheet 3, 4, and 5 for each student
- a copy of *A Missing Link* Worksheet 6 for each student
- access to the internet, or local library
- chart paper, markers, pencils, pencil crayons

PROCEDURE:
*This lesson can be done as one long lesson, or be divided into three shorter lessons.

1. Using Worksheets 1 and 2, do a shared reading activity with the students. This will allow for reading practice and learning how to break down word parts in order to read the larger words in the text. Along with the content, discussion of certain vocabulary words would be of benefit for students to fully understand the passage.

    Some interesting vocabulary words to focus on are: ***environment, chlorophyll, castings, hydrated, natural, carbon dioxide, nutrients, liquid, urine, recyclers, oxygen, digest, absorb, returned, breathes/breathing***

2. Using Worksheet 3, do a shared reading activity with the students. Along with the content, discussion of certain vocabulary words would be of benefit for students to fully understand the passage.

    Some interesting vocabulary words to focus on are: ***exchange, beaver dam, insects, nectar, re-create, gnaw, stumps, pollen, provide***

3. Give students Worksheets 4 and 5. They can visit the local library or access the internet to gather information to help them answer the questions on Worksheets 4 and 5. Upon completion, students can participate in a 'turn and talk' activity with a partner, where they share the information they have acquired on Worksheet 5.

4. Divide students into pairs and give them Worksheet 6. They will engage in a **think, pair, share** activity to discuss what the world would be like without certain living things in it. They will choose one of their ideas to illustrate the impact of its loss on the environment. Follow up with a large group discussion by asking some students to share their ideas. Pose this question to the group: Who would be affected by the loss, and how?

DIFFERENTIATION:
Slower learners may benefit by a reduction in expectations by the elimination of Worksheet 5.

For enrichment, faster learners could make a list of jobs or hobbies that require knowledge about the needs of living things, and then choose one to illustrate and write about.

Processing Needs

You have learned a lot about the basic needs of living things. You know that all living things need water, food, air, and shelter to stay alive.

When a basic need is met, it gets used by a living thing, and it returns to the environment in a different form. Let's learn more about this idea!

Question: What happens to the air that a human being uses to meet its need of breathing?

Answer: The oxygen that a human breathes in is used to bring air into the body. When a human breathes out, the air comes out into the environment as carbon dioxide.

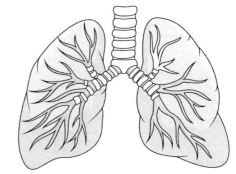

Question: What happens to the air that a plant uses to meet its need of breathing?

Answer: The carbon dioxide that a plant breathes in is used to bring air into it which is mixed with its chlorophyll to make plant food. When the plant breathes out, the air comes out into the environment as oxygen.

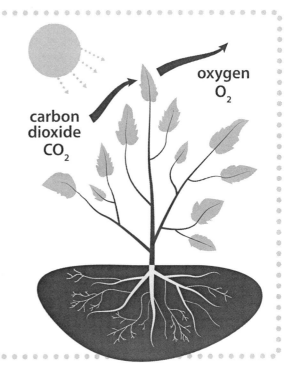

Worksheet 2 Name: _____

Question: What happens to vegetable scraps that an earth worm uses to meet its need of food?

Answer: When an earthworm eats and then digests its food, its castings are left behind, which add nutrients to plants in the soil.

Question: What happens to the liquid that wild animals use to meet their need for water?

Answer: When wild animals drink water, they absorb what they need to be hydrated, and then it is returned to the Earth as urine.

Did you know?

Red-bellied Woodpeckers meet their need of shelter by making nests inside dead trees. They chip away the wood to make a hole in the tree. Red-bellied Woodpeckers are natural recyclers in the environment because they use the left over woodchips by making a bed with them to lay their eggs on.

Worksheet 3 Name: _____

Give and Take

While living things get their needs met, sometimes this action leaves behind something that another living thing could also use to meet its needs.

You have learned that as humans breathe out carbon dioxide, it is taken in by plants, which breathe out oxygen that humans take in. This is an exchange of needs.

Did you know?

Beavers use their strong teeth to gnaw down trees which they need in order to make a beaver dam. The left over tree stumps can be useful to other living things. For example, a chipmunk will meet its need for shelter by making its home in a tree stump.

Insects and plants help each other meet their needs too. When insects land on plants, they carry and spread pollen that plants need to re-create. Plants provide shelter and food for some insects. For example, a bee gets the nectar it needs from a plant to make honey and the bee spreads the pollen of plants as it flies from flower to flower.

Worksheet 4   Name: _____

Visit a library or use a computer to help you find out the answers to these questions about what some living things provide for each other.

Question: What does a cow provide for other living things?

_____
_____
_____
_____

What needs does this meet?

_____
_____
_____

Question: What does a maple tree provide for other living things?

_____
_____
_____

What needs does this meet?

_____
_____
_____

Worksheet 5   Name: _____

It is your turn to ask some question of your own about what some living things provide for each other. Don't forget to find out the answers to your questions!

Question: _____
_____

What does a _____ provide for other living things?
_____
_____

What needs does this meet?
_____
_____

Question: _____
_____

What does a _____ provide for other living things?
_____
_____

What needs does this meet?
_____
_____